T0269103

This fifth volume of Colin Ronan's abridgement of Joseph Needhams's monumental work is concerned with the staggering engineering feats made in early and medieval China. No other country did more in civil engineering, both as to scale and skill, than China. The book opens with an account of the road system, which compared not unfavourably with that of the Roman Empire. Naturally, the Great Wall of China is covered in some detail within the social context of walls, which mark, more than any other structure, the basic features of the Chinese communities. The Chinese genius of town planning and achievements in hydraulic engineering are covered in later chapters. This book concludes the abridgement of the Chinese engineering, and provides many clues as to the influence of Chinese innovation on Western engineering trends.

THE SHORTER
SCIENCE AND CIVILISATION IN CHINA

COLIN A. RONAN
Lately of the Needham Research Institute

The Shorter Science and Civilisation in China

AN ABRIDGEMENT OF

JOSEPH NEEDHAM'S ORIGINAL TEXT

Volume 5

THE FIRST SECTION OF VOLUME IV, PART 3
THE FINAL SECTION OF VOLUME IV, PART 2 OF
THE MAJOR SERIES

ROADS
WALLS & THE GREAT WALL
BUILDING TECHNOLOGY
BRIDGES
HYDRAULIC ENGINEERING I – CONTROL,
 CONSTRUCTION & MAINTENANCE OF
 WATERWAYS
HYDRAULIC ENGINEERING II – WATER-RAISING
 MACHINERY & WATER AS A POWER SOURCE

CAMBRIDGE
UNIVERSITY PRESS

CAMBRIDGE UNIVERSITY PRESS
Cambridge, New York, Melbourne, Madrid, Cape Town, Singapore, São Paulo

Cambridge University Press
The Edinburgh Building, Cambridge CB2 2RU, UK

Published in the United States of America by Cambridge University Press, New York

www.cambridge.org
Information on this title: www.cambridge.org/9780521462143

© Cambridge University Press 1995

This publication is in copyright. Subject to statutory exception
and to the provisions of relevant collective licensing agreements,
no reproduction of any part may take place without
the written permission of Cambridge University Press.

First published 1995

A catalogue record for this publication is available from the British Library

Library of Congress Cataloguing in Publication data

(Revised for volume 5)
Needham, Joseph, 1900–
The Shorter Science and Civilisation in China.
Volume 2 contains Volume III and a section of Volume
IV, part 1 of the major series.
Volume 3 contains a section of Volume IV, part 1,
and a section of Volume IV, part 3 of the major series;
Volume 5 contains Volume IV, part 3 and the final
section of Volume IV, part 2 of the major series.
Includes bibliographies and indexes.
1. China – Civilization.   2. China – Intellectual
life.   3. Science – China – History.   I. Ronan, Colin A.
DS721.N392   951   77–82513

ISBN-13 978-0-521-46214-3 hardback
ISBN-10 0-521-46214-2 hardback

ISBN-13 978-0-521-46773-5 paperback
ISBN-10 0-521-46773-X paperback

Transferred to digital printing 2006

CONTENTS

ILLUSTRATIONS

TABLES

PREFACE

In this, the fifth volume of Dr Joseph Needham's *Science and Civilisation in China*, we look at various aspects of civil engineering in ancient and medieval China. This involves covering the whole civil engineering section of volume IV, part 3 of Dr Needham's original work, and also that section of his volume IV, part 2, concerned with hydraulic machinery. The text thus covers the vast Chinese road system, walls and the Great Wall, the great Chinese achievements in bridge building, their unique system of canals and inland waterways, together with their machinery for utilising the power of water.

As in previous volumes in the abridgement, I am indebted to Joseph Needham for his encouragement and help; as always, his advice has been invaluable. In keeping with previous volumes, this is no new edition, but some changes are made by use of the Pinyin romanisation. For ease of reading the text, only Pinyin is given, except in those instances where Chinese words themselves are being discussed; on these occasions the Pinyin is accompanied by the modified Wade–Giles system in square brackets. Moreover, we have followed the procedure in the *Cambridge Encyclopedia of China* for place names; that is to give Pinyin only except for such well-known romanisations as Canton and Yangtze. Pinyin is used for all personal and place names in the index (except those where Pinyin and the modified Wade–Giles are the same), but the modified Wade–Giles is also given in square brackets, thus enabling readers who wish to do so to find specific references in Dr Needham's original volumes.

My warmest thanks are due to Mr John Moffett, Librarian of the Needham Research Institute, and to Ms Corrine Richeux, Dr Needham's assistant there, for their kind and unstinting help in various matters. I am also much indebted to Dr Simon Mitton and Ms Fiona Thomson of Cambridge University Press for their ever ready help, as well as to Mrs Irene Pizzie and Mrs Tracey Humphries for copy editing and checking the Pinyin transliterations, respectively. I am also indebted to Ms Liz Granger for compiling the index.

The preparation of this volume has only been possible due to the receipt of a grant from the Chiang Ching-kuo Foundation for International Scholarly Exchange in Taiwan. For this very valuable help I am most grateful.

Hastings, East Sussex   Colin A. Ronan

1

Roads

No ancient country in the world did more in civil engineering, both as to scale and skill, than China, and this is particularly so when it comes to hydraulic engineering in all its aspects. Yet the Chinese did not neglect roads, and the system they developed compares not unfavourably with that of the Roman Empire when we consider it as only part of their system of communications. After all, as Adam Smith wrote in AD 1776, 'Good roads, canals and navigable rivers by diminishing the expense of carriage, put the remote parts of a country more nearly upon a level with those in the neighbourhood of the town. They are upon that account the greatest of improvements.' When one remembers that the Chinese Empire covered an area of almost 4 million square kilometres (some 1.5 million square miles), it becomes clear that a total of over 35 400 kilometres (22 000 miles) of specially made roads by AD 190 during the Later Han dynasty was no mean achievement.

The mode of construction of Roman roads is clearly shown in remains that still exist, and it is also well documented. In a ditch some 1.5 to 1.8 metres deep, a bed of stones was laid, then a layer of rubble and chippings, followed by sand, or gravel, or broken pottery and bricks locked in place by a lime cement. Flat stone slabs made up the final surface. Kerbs were often provided. Sometimes, the lower layers were extended sideways, with a ditch on either side, and on occasions the road was accompanied by a drainage channel of substantial size, while retaining walls might be built along the sides of steep slopes. The Romans also used graded earth tracks and side-roads with a gravelled surface.

It has often been said that roads in the Roman style resembled, to some extent, a series of walls lying horizontally, and these have often been highly praised. Yet, as the historian Lefebvre des Noëttes pointed out, they were, in truth, primitive and ill-suited to their purpose. Allowing nothing for expansion and contraction due to temperature, frost fissures and unequal drainage, they depended on thickness and rigidity. Yet the more successful methods of later times, culminating in the second decade of the nineteenth century with the compacted chips of John McAdam, and their subsequent developments, all depend on thinness and elasticity. These appear to be medieval in origin, but Chinese roads of similar light and elastic type long preceded them, as we shall see.

NATURE AND EXPANSION OF THE NETWORK

Gazing down on the Old World during the few centuries before and after the turn of our era, one could have seen, as in a slow-motion video, the appearance and radiation of two branching systems of highway communications: one springing from the western coast of the ancient Italian peninsula, the other near the great bend of the Yellow River where it swings round the Shanxi mountains to flow eastwards to the Yellow Sea. Should the Romans have ever succeeded in conquering the Parthians and Persians, the two road systems might have met, perhaps somewhere west of Xinjiang, but this was not to be. The octopus-like arms expanded independently, each in a world of its own, their builders troubled only occasionally by the vaguest rumours of another system too far away to matter.

There is a curious parallel between the Roman and Chinese systems in that both, after the third century AD, fell into a long period of decay, but while Europe became parcelled out into feudal kingdoms and domains with poor communications except by sea, the role of the Chinese highways passed over to an immense system of navigable rivers and artificial waterways, leaving only the mountain roads to continue their age-old function. For, as always to be expected in a feudal-bureaucratic society, the central government concerned itself with the construction and maintenance of the most satisfactory routes of communication.

This may be illustrated by some of the oldest records of road building in the Chinese culture-area which have come down to us. A verse in the *Shi Jing* (Book of Odes) expresses admiration of the roads in the neighbourhood of the capital of the Zhou State:

> The roads of Zhou are (smooth) as a whetstone,
> Straight as an arrow('s flight);
> Ways where the lords and officials pass,
> Ways where the common people look on.

This folk-song is considered rather ancient, perhaps of the ninth century BC, in the Western Zhou period. When we come to the *Zhou Li* (Record of Institutions (lit. Rites) of the Zhou Dynasty), that second century BC compilation of the ideal structure of the feudal-bureaucratic State, we have much more detailed information on the technical terms for roads. Interestingly, the Record seems to incorporate two distinct traditions, probably from different earlier feudal States. In the entry for the Si Xian (Director of Communications), we read:

> He studies the maps of the nine provinces in order to obtain a perfect knowledge of the mountains, forests, lakes, rivers and marshes, and to understand the (natural) routes of communication.
>
> [Commentary. When mountains and forests present obstacles, he cuts through them. When rivers and lakes offer impediment, he bridges them.]
>
> He lays out the five kinds of canal and the five kinds of road, planting trees and hedges along them for defence. All (special points, passes and junctions) have guard-posts, and he knows the ways and roads that lead to them.
>
> [Commentary. The five kinds of canal are *sui* (ditches), *kou* [*khou*] (conduits), *xue* [*hsüeh*] or *xu* [*hsü*] (small canals), *gui* [*kuei*] or *guai* [*kuai*] (medium canals), and *chuan* [*chhuan*] (great canals). The five kinds of road (*tu* [*thu*]) are *jing* [*ching*] (paths or ways), *zhen* [*chen*] (larger, paved ways), *tu* [*thu*] (one-width roads), *dao* [*tao*] (two-width roads), and *lu* (three-width roads).]
>
> If there is alarm in the empire he fortifies the roads and difficult points, halts wanderers, and guards the positions with his men, letting past the barriers only those with the imperial seal.

The systematisation of the capacities of roads and canals, doubtless largely schematic, appears in the passages devoted to the Sui Ren (Grand Extensioner, or Minister of Agriculture):

> This is how he organises the countryside. Between each farm there is a ditch (*sui*) with a path (*jing* [*ching*]) along it. Past every ten farms there runs a conduit (*gou* [*kou*]) with a way (*zhen* [*chen*]) alongside. Past every hundred farms there runs a small canal (*xue* [*hsüeh*]) with a one-width road (*tu* [*thu*]) accompanying it. Past every thousand farms there runs a medium-sized canal (*gui* [*kuei*]) with a two-width road (*dao* [*tao*]) along its bank. Past every ten thousand farms there runs a large canal (*chuan* [*chhuan*]) with a three-width road (*lu*) at its side. Such are the communications in the imperial domains.
>
> [Commentary. The five grades of roads are all to connect the country and the capital for carriages and pedestrians. (Apart from men) paths will take only horses and oxen, the wider (paved) ways will take large hand-carts, a one-width road will take a single chariot, a two-width road will take two abreast, and a three-width road will take three abreast. One may make the country roads the same width as the ring-roads of cities.]

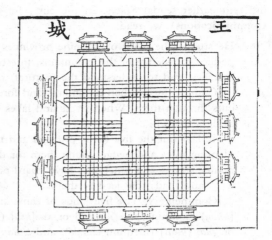

Fig. 389. Diagram of an idealised imperial or princely city, with its thoroughfares. From the *San Li Tu* (Illustrations (diagrams) of the Three Rituals).

Now we know what was meant by a 'two-width road'. But another text in the same book has more spacious ideas. Under the heading of Jiang Ren (Master-Builders) we find that in the capital the main streets (*jing tu* [*ching thu*]) are to carry nine chariots abreast, the ring-roads (*huan tu* [*huan thu*]) are to carry seven, and the country roads (presumably imperial highways, *ye tu* [*yeh thu*]) are to carry five (Fig. 389). Furthermore, capitals of feudal princes are to have their main streets of the seven-width grade, their ring-roads five-width, and their approach roads three-width. Other cities must not exceed the five-width grade for their broadest streets, with all their other roads at the three-width level. Perhaps there is no discrepancy if the grandeur of the Zhou (or Han) capital is at issue only in this second text.

During the Warring States period, there was much road-building activity, both for military and commercial purposes. The State of Qin, however, as we shall now see, had been particularly busy, and the works achieved may well have been a great factor in its success. As soon as the whole empire was for the first time united under Qin Shi Huang Di in 221 BC, the new ruler embarked upon his celebrated policy of standardisation of measurements, and fixed, among other things, the gauge of chariot-wheels. In 220 BC, Qin Shi Huang Di made a tour of inspection in Gansu and Shaanxi, after which he ordered the construction of a vast set of arterial post-roads, 'speed-ways' (*chi dao* [*chhih tao*]) or 'straight-ways' (*zhi dao* [*chih tao*]) radiating from the capital at Chang'an (near modern Xi'an), especially to the north, north-east, east and south-east.

Though contemporary descriptions are not available, it is worth giving one from only a few years afterwards. In about 178 BC, Jia Shan, one of

Emperor Wen Di's counsellors, presented an essay analysing the causes of good government and civil confusion, particularly criticising Qin Shi Huang Di. After decrying the luxury of the palaces built at Xianyang, he continued:

> He also ordered the building of post-roads all over the Empire . . . so that all was made accessible. These highways were 50 paces wide, and a tree was planted every 9 metres along them. The road was made very thick and firm at the edge, and tamped with metal rammers. The planting of the green pine-trees was what gave beauty to the roads. Yet all this was done (only) so that Qin Shi Huang Di's successors (on the throne) should not have to take circuitous routes.

Later commentators were a little puzzled by the statement about the structure of the roads, some thinking they were lined by walls on each side like raised causeways, others that the tamping referred simply to the consolidation of the edges, especially where there was an embankment. That little trace of these roads remained in later ages presumably implies that they were less massively built than the Roman roads. Yet if they consisted chiefly of rubble and gravel tamped down in the manner of *pisé* walls (see p. 26), they were more elastic and much more modern in conception. Such 'water-bound macadam' was in fact the traditional material of Chinese highways in all periods.

As for the width, it is generally agreed that the '50 paces' of the *Qian Han Shu* (History of the Former Han Dynasty), compiled from about AD 65 onwards, was a scribal error for '50 feet' (9 metres), so that the imperial highways would have been approximately nine-width roads equivalent to the broadest described in the *Record of the Institutions of the Zhou Dynasty*. They were thus larger than most of the Roman roads.

A few comments on the map of the imperial highways shown in Fig. 390 may now usefully be made. A more easterly centre, Sanchuan (6) in the neighbourhood of modern Luoyang (Henan province), was chosen as the hub of the system, the road from nearby Chang'an (1) negotiating the Hangu pass (12) much as the railway does today. It then splits, one arm going northwards to Ji (7) near modern Bejing, another north-east to Linzi or Qi (8), following the old course of the Yellow River for part of the way. The third and longest leads south-eastwards to Pei (10) (north of present Pei), from whence it wends its way gradually southwards over the Yangtze close to modern Nanjing and on to Wu (11), the capital of the former Wu State. Nearly as long as the south-eastern road is the southern one, which goes over the mountains by way of the Wu Pass (13), thence to Nanyang (14) and southwards, crossing the Han river and then the Yangtze near the Dongting Lake. From Changsha (16), it followed the Xiang River valley,

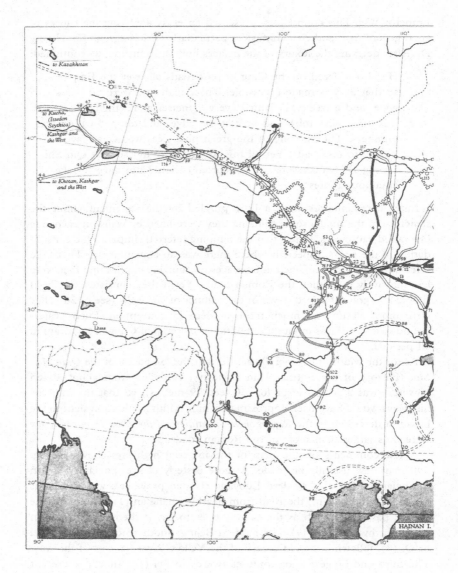

ending its journey at Lingling (18). Though now bearing somewhat east-wards, this was no mistake, for, as we shall see in chapter 5, the upper waters of the Xiang were made in Qin times to connect with the upper waters of the West River of Guangdong, thus permitting the transport of arms and supplies for the conquest of the Cantonese State of Nan Yue.

Something must now be said of the Great North Road, the only road for

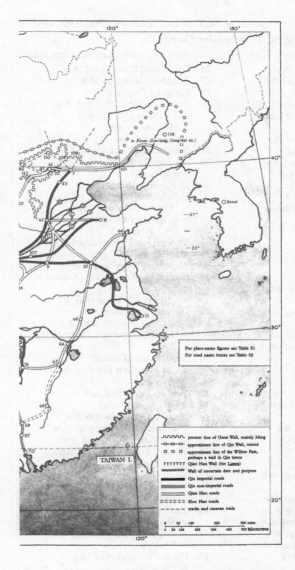

Fig. 390. Map
of road
communications
in ancient
China, and of
the lines of the
great defensive
walls.

which we have any details concerning its construction. In 212 BC, Meng
Tian, one of the Qin First Emperor's most important generals, whose name
will always be connected with the Great Wall (see p. 41), was ordered to
build a road from Xianyang (2), the imperial capital, up through the Ganquan
districts (3, 4) and out through the Wall, striding across the Ordos Desert
plateau of Inner Mongolia to the northernmost line of the Yellow River.

Table 51. *Place names for the maps of road communications (Fig. 390) and civil engineering works (Fig. 464) in ancient China.*

1 Chang'an [Chang-an] 長安
  = Xi'an [Hsi-an (Sian)] 西安
2 Xianyang [Hsien-yang] 咸陽
  = Weicheng [Wei-chhêng] 渭城
3 Ganquan Shan [Kan-chhüan Shan] 甘泉山
  = Shunhua [Shun-hua] 醇化
  = Yunyang [Yün-yang] 雲陽
4 Ganquan [Kan-chhüan] 甘泉
5 Jiuyuan [Chiu-yuan] 九原
  = Wuyuan [Wu-yuan]
6 Luoyang [Lo-yang] 洛陽
  = Sanchuan [San-chhuan] 三川
7 Yan [Yen] 燕
  = Ji [Chi] 薊
  = Beijing 北京
8 Linzi [Lin-tzu] 臨菑
  = Qi [Chhi] 齊
9 Kaifeng [Khai-fêng]
  = Chenliu [Chhen-liu] 陳留
  = Daliang [Ta-liang] 大梁
  = Bianjing [Pien-ching] 汴京
10 Pei [Phei] 沛
  = Peixian [Phei-hsien] 沛縣
11 Suzhou [Su-chou] (Suchow) 蘇州
  = Wu [Wu] 吳
  = Huiji [Hui-chi] 會稽
12 Hangu Guan [Han-ku Kuan] 函谷關 (pass)
13 Wu Guan [Wu Kuan] 武關 (pass)
14 Nanyang [Nan-yang] 南陽
  = Wan [Wan] 宛
15 Nan [Nan] 南
  = Nanjun [Nan-chün] 南郡
  = Jiangling [Chiangling] 江陵
  = Ying [Ying] 郢
  = Linjiang [Lin-chiang] 臨江
  = Jingzhou [Ching-chou] 荊州
16 Changsha [Chhang-sha] 長沙
17 Hengyang [Hêng-yang] 衡陽
  = Hengshan [Hêng-shan] 衡山
18 Lingling [Ling-ling] 零陵
19 Guilin [Kuei-lin] 桂林
20 Xiang [Hsiang] 象
  = Cangwu [Tshang-wu] 蒼梧
  = Wuzhou [Wu-chou] 梧州
21 Wugong [Wu-kung] 武功
22 Fufeng [Fu-fêng] 扶風
23 Baoji [Pao-chi] 寶雞
  = Chencang [Chhen-tshang] 陳倉

24 Tianshui [Thien-shui] 天水
25 Longxi [Lung-hsi] 隴西
26 Dingxi [Ting-hsi] 定西
27 Jincheng [Chin-chhêng] 金城
  = Lanzhou [Lan-chou] (Lanchow) 蘭州
  = Gaolan [Kao-lan] 皋
28 Yongdeng [Yung-têng] 永登
29 Wushao Ling [Wu-shao Ling] 烏鞘嶺 (pass)
30 Liangzhou [Liang-chou] (Liangchow) 涼州
  = Wuwei [Wu-wei] 武威
  = Sera Metropolis (by mistake for Chang'an)
31 Yongchang [Yung-chhang] in Gansu province 永昌
32 Shandan [Shan-tan] 山丹
33 Ganzhou [Kan-chou] (Kanchow) 甘州
  = Zhangye [Chang-yeh] 張掖
34 Gaotai [Kao-thai] 高臺
35 Jiuquan [Chiu-chhüan] 酒泉
  = Suzhou [Su-chou] (Suchow)] 肅州
36 Jiayu Guan [Chia-yü Kuan] 嘉峪關 (Western Gate of the Great Wall)
37 Yumen [Yü-mên] 玉門
38 Anxi [An-hsi] 安西
  = Guazhou [Kua-chou] 瓜州
39 Dunhuang [Tun-huang] 敦煌
  = Shazhou [Sha-chou] 沙州
40 Yumen Guan [Yü-mên Kuan] 玉門關 (Jade Gate)
41 Yiwu [I-wu] 伊吾
  = Hami [Hami] 哈密
42 Loulan [Lou-lan] 樓蘭
43 Yuni [Yü-ni] 窳匿
  = Shanshan [Shan-shan] 鄯善
  = Erqiang [Erh-chhiang] 婼羌
  = Charlik
44 Qiemo [Chhieh-mo] 且末
  = Cherchen
45 Gaochang [Kao-chhang] 高昌
  = Shanshan [Shan-shan] 鄯善
  = Karakhoja
  = Turfan
46 Jiaohe [Chiao-ho] 交河
  = Piala
  = Yarkhoto

Table 51. (*cont.*)

47	Yanqi [Yen-chhi] 焉耆 = Karashahr
48	Weili [Wei-li] 尉犁 = Kalgaman
49	Beidi [Pei-ti] 北地 = Ningxian [Ning-hsien] 寧縣
50	Anding [An-ting] 安定 = Pingliang [Phing-liang] 平涼
51	Yong [Yung] 雍 = Fengxiang [Fêng-hsiang] 鳳翔
52	Xiao Guan [Hsiao Kuan] 蕭關 (Xiao pass)
53	Huizhong Gong [Hui-chung Kung] 回中宮 = Guyuan [Ku-yuan] 固原
54	Liyang [Li-yang] 櫟(or 櫟)陽
55	Shangjun [Shang-chün] 上郡
56	Yulin [Yü-lin] 榆林
57	Huayin [Hua-yin] 華陰
58	Hongnong [Hung-nung] 弘(or 宏)農 = Guolüe [Kuo-lüeh] 虢略
59	Hedong [Ho-tung] 河東
60	Jinyang [Chin-yang] 晉陽 = Taiyuan [Thai-yuan] 太原
61	Dai [Tai] 代 = Daijun [Tai-chün] 代郡
62	Handan [Han-tan] 邯鄲
63	Zhongshan [Chung-shan] 中山
64	Langya [Lang-ya] 琅邪 = Langye [Lang-yeh] 琅琊 (the guantai [kuan-thai] 觀臺, observation terrace)
65	Lujiang [Lu-chiang] 廬江
66	Jiujiang [Chiu-chiang] 九江
67	Qingjiang [Chhing-chiang] 清江
68	Ganxian [Gan(-hsien] 贛(縣)
69	Qujiang (Gugong) [Chhü-chiang (Kukong)] 曲江 = Shaoguan [Shao-kuan] 韶關 = Shaozhou [Shao-chou] 韶州
70	Nanhai [Nan-hai] 南海 = Guangzhou [Kuang-chou (Canton)] 廣州
71	Xiangyang [Hsiang-yang] 襄陽
72	Mianxian [Mien-hsien] 沔縣
73	Baocheng [Pao-chhêng] 襄城
74	Hanzhong [Han-chung] 漢中 = Nanzheng [Nan-chêng] 南鄭
75	Fengxian [Fêng-hsien] 鳳縣 = Shuangshipu [Shuang-shih- phu] 雙石鋪
76	Liuba [Liu-pa] 留壩

77	Meixian [Mei-hsien] 郿縣
78	Zhouzhi [Chou-chih] 盩屋
79	Ningqiang [Ning-chhiang] 寧羌 (or 強)
80	Zhaohua [Chao-hua] 昭化
81	Jianmen Guan [Chien-mên Kuan] 劍門關 (Sword-gate Pass)
82	Mianyang [Mien-yang] 綿陽
83	Shu [Shu] 蜀 = Chengdu [Chhêng-tu] 成都
84	Ba [Pa] 巴 = Chongqing [Chung-chhing] (Chungking) 重慶
85	Bayu Guan [Pa-yü Kuan] 巴峪關 (pass)
86	Pingcheng [Phing-chhêng] 平城 = Datong [Ta-thung (Tatung)] 大同
87	Feihu Kou [Fei-hu Khou] 飛狐口 (Flying-fox Pass)
88	Zigui [Tzu-kuei] 秭歸
89	Podao [Pho-tao] 焚道 = Yibin [I-pin] 宜賓
90	Dian [Tien] 滇 = Dianchi [Tien-chhih] 滇池 = Kunming [Khun-ming] 昆明
91	Yeyu [Yeh-yü] 葉(or 楪)榆 = Dali [Ta-li] 大理
92	Zangke [Tsang-kho] 牂(or 牂)柯 (or 柯 or 柯)
93	Yuesui [Yüeh-sui] (越巂 or 嶲) = Qiongdu [Chhiung-tu] 邛都
94	Hepu [Ho-phu] 合浦 = Leizhou [Lei-chou] 雷州 = Haikang [Hai-khang] 海康
95	Jiaozhou [Chiao-chou] 交州 = Hanoi
96	Hanguang [Han-kuang] 洽光 (or 洸)
97	Yingde [Ying-tê] 英德 = Zhenyang [Chên-yang] 湞陽
98	Guiyang [Old Kuei-yang] 桂陽 = Yuanling [Yuan-ling] 沅陵
99	Juyan [Chü-yen] 居延 = Edsin (or Etsin) Gol
100	Yongchang 永昌 (in Yunnan) = Baoshan [Pao-shan] 保山
101	Shanhai Guan [Shan-hai Kuan] 山海關 (Eastern Gate of the Great Wall)
102	Yelang [Yeh-lang] 夜郎 = Tongzi [Thung-tzu] 桐梓

Table 51. (*cont.*)

103 Langzhou [Lang-chou] 郎州
    = Zunyi [Tsun-I] 遵義
	111 Zijing Guan [Tzu-ching Kuan]
	    紫荊關 (pass)

104 Yizhou [I-chou] 益州
    = Chengjiang [Chhêng-chiang] 澂江
	112 Pingxing Guan [Phing-hsing
	    Kuan] 平型關 (pass)

105 Jumi [Chü-mi] 且彌
    = Zhenxi [Chen-hsi] 鎮西
    = Balikun [Pa-li-khun] 巴理坤
    = Barkol
	113 Yanmen [Yen-mên] 雁門
	    = Youyu [Yu-yü] 右玉
	114 Ningxia [Ning-hsia] 寧夏
	    = Yinchuan [Yin-chhuan] 銀川

106 Yizhi [I-chih] 移支
    = Dihua [Ti-hua] 迪化
    = Urumchi
	115 Xining [Hsi-ning] 西寧
	116 Yang Guan [Yang Kuan] 陽關
	    (gate)

107 Rongyang [Jung-yang] 滎陽
    = Zhengzhou [Chêng-chou
    (Chêngchow)] 鄭州
	117 Gaoque [Kao-chhüeh] 高闕
	118 Shenyang [Shen-yang] 瀋陽
	    = Mukden

108 Gubeikou [Ku-pei-khou] 古北口
109 Nankou [Nan-khou] 南口
    = Juyong Guan [Chü-yung
    Kuan] 居庸關 (gate)
	119 Lintao [Lin-thao] 臨洮
	    = Minzhou [Min-chou] 岷州
	120 Suzhou [Su-chou (Suchow)]
	    宿州

110 Zhangjiakou [Chang-chia-khou]
    張家口
    = Wanquan [Wan-chhüan] 萬全
    = Kalgan
	121 Shexian [Shê-hsien] 葉縣
	122 Shouzhou [Shou-chou] 壽州
	123 Daming [Ta-ming] 大名
	124 Linqing [Lin-chhing] 臨清

Anding, 50	Dianchi, 90
Anxi, 38	Dihua, 106
Ba, 84	Dingxi, 26
Balkun, 105	Dunhuang, 39
Baocheng, 73	Edsin Gol, 99
Baoji, 23	Erqiang, 43
Baoshan, 100	Feihu Kou, 87
Barkol, 105	Fengxian, 75
Bayu Guan, 85	Fengxiang, 51
Beidi, 49	Fufeng, 22
Beijing, 7	Ganquan, 4
Bianjing, 9	Ganquan Shan, 3
Cangwu, 20	Ganxian [Gan-hsien], 68
Chang'an, 1	Ganzhou (Kanchow), 33
Changsha, 16	Gaochang, 45
Charlik, 43	Gaolan, 27
Chencang, 23	Gaoque, 117
Chengdu, 83	Gaotai, 34
Chengjiang, 104	Guangzhou (Canton), 70
Chenliu, 9	Guazhou, 38
Cherchen, 44	Gubeikou, 108
Chongqing (Chungking), 84	Guilin, 19
Dai, 61	Guiyang, 98
Daijun, 61	Guolüe, 58
Dailing, 8	Guyuan, 53
Dali, 91	Haikang, 94
Daming, 123	Hami, 41
Datong (Tatung), 86	Handan, 62
Dian, 90	Hangu Guan, 12

Table 51. (*cont.*)

Hanguang, 96	Nanjun, 15
Hanzhong, 74	Nankou, 109
Hedong, 59	Nanyang, 14
Hengshan, 17	Nanzheng, 74
Hengyang, 17	Ningqiang, 79
Hepu, 94	Ningxia, 114
Hongnong, 58	Ningxian, 49
Huayin, 57	Pei, 10
Huiji, 11	Peixian, 10
Huizhong Gong, 53	Piala, 46
Ji, 7	Pingcheng, 86
Jiangling, 15	Pingliang, 50
Jianmen Guan, 81	Pingxing Guan, 112
Jiaohe, 46	Podao, 89
Jiaozhou, 95	Qi, 8
Jiayu Guan, 36	Qiemi, 105
Jincheng, 27	Qiemo, 44
Jingzhou, 15	Qingjiang, 67
Jinyang, 60	Qiongdu, 93
Jiuchuan, 35	Qujiang (Gugong), 69
Jiujiang, 66	Rongyang, 107
Jiuyuan, 5	Sanchuan, 6
Juyan, 99	Shandan, 32
Juyong Guan, 109	Shangjun, 55
Kaifeng, 8	Shanhai Guan, 101
Kalgaman, 48	Shanshan, 43, 45
Karakhoja, 45	Shenyang (Mukden), 118
Karashahr, 47	Shaoguan, 69
Kunming, 91	Shaozhou, 69
Langya, 64	Shazhou, 39
Langye, 64	Shexian, 121
Langzhou, 103	Shouzhou, 122
Lanzhou (Lanchow), 27	Shu, 83
Leizhou, 94	Shuangshipu, 75
Liangzhou (Liangchow), 30	Sunhua, 3
Lingling, 18	Suzhou, 35
Linjiang, 15	Suzhou (Suchow), 11
Linqing, 124	Suzhou (Suchow), 120
Lintao, 119	Taiyuan, 60
Linzi, 8	Tianshui, 24
Liuba, 76	Tongzi, 102
Liyang, 54	Turfan, 45
Longxi, 25	Wan, 14
Loulan, 42	Wanquan, 110
Lujiang, 65	Weicheng, 2
Luoyang, 6	Weili, 48
Meixian, 77	Wu, 11
Mianxian, 72	Wu Guan, 13
Mianyang, 82	Wugong, 21
Minzhou, 119	Wushao Ling, 29
Nan, 15	Wuwei, 30
Nanhai, 70	Wuyuan, 5

Table 51. (*cont.*)

Wuzhou, 20	Yongchang (in Yunnan), 100
Xian, 1	Yongdeng, 28
Xiang, 20	Youyu, 113
Xiangyang, 71	Yuanling, 98
Xiao Guan, 52	Yuesui, 93
Xining, 115	Yulin, 56
Yan (modern Beijing), 7	Yumen, 37
Yang Guan, 116	Yumen Guan, 40
Yanmen, 113	Yuni, 43
Yanqi, 47	Yunyang, 3
Yarkhoto, 46	Zangke, 92
Yelang, 102	Zhangjiakou, 110
Yeyu, 91	Zhangye, 33
Yibin, 89	Zhaohua, 80
Yinchuan, 114	Zhengzhou, 107
Ying, 15	Zhenxi, 105
Yingde, 97	Zhenyang, 97
Yiwu, 41	Zhongshan, 63
Yizhi (Urumchi), 106	Zhouzhi, 78
Yizhou, 104	Zigui, 88
Yong, 51	Zijing Guan, 111
Yongchang (in Gansu), 31	Zunyi, 103

Here, somewhere not far west of the modern steel city of Baotou, was a fortified outpost called Jiuyuan (5), probably a vantage-point for observing the affairs of the Huns and other nomadic peoples, and doubtless also a trading-post for their products. The texts expressly state that this road was carried in a straight line through the mountains by cuttings and across valleys by embankments.

It is interesting to assess the total length of these imperial highways which fanned out from the metropolitan region of Guanzhong in the third century BC; they amount to no less than 6840 kilometres (4250 miles) – not very unlike the distance of some 6000 kilometres (3740 miles) which the eighteenth-century historian, Edward Gibbon, estimated as the length of the first-rate Roman road running from the Antonine Wall in Scotland to Jerusalem.

There is evidence that the imperial highways system was only part of the road-building activity of the Qin. The city Yong (51), its former capital, was certainly connected with Chang'an (1), and by the time the dynasty fell in 209 BC, some kind of road existed connecting Hengyang (17) in the south with Canton (70) on the southern coast. This must have penetrated gorges of the Nanling mountains, and in later centuries was to become a very important north–south route between Canton and Beijing, though for a long time the bulk of the traffic went round by the canal near Lingling (18). But

Table 52. *Highway names on the map of road communications in ancient China (Fig. 390).*

A	Old Silk Road		
B	Lianyun Dao [Lien-yün Tao]	連雲道	(Linked-Cloud Road)
C	Baoxie Dao [Bao-hsieh Tao]	襃斜道	(Baoxie Road)
	= Beizhan Lu [Pie-chan Lu]	北棧路	(Northern Trestle and
	= The New Road		Gallery Road)
D	Tangluo Dao [Tahang-lo Tao]	儻駱道	(Luo Valley Road)
	= Luogu Dao [Lo-ku Tao]	駱谷道	
E	Ziwu Dao [Tzu-wu Tao]	子午道	(North–South Road)
F	Chencang Dao [Chhen-tshang Tao]	陳倉道	(Chen's Granary Road)
	= The Old Road		
G	Jinniu Dao [Chin-niu Tao]	金牛道	(Road of the Golden Oxen)
	= Shiniu Dao [Shih-niu Tao]	石牛道	(Road of the Stone Oxen)
H	Micang Dao [Mi-tshang Tao]	米倉道	(Rice Granary Road)
J	Feihu Dao [Fei-hu Tao]	飛狐道	(Flying Fox Road)
K	Wuchi Dao [Wu-chhih Tao]	五尺道	(Five-Foot Way)
L	Tianshan Bei Lu [Thien-Shan Pei Lu]	天山北路	(Road north of the Tianshan)
M	Tianshan Nan Lu [Thien-Shan Nan Lu]	天山南路	(Road south of the Tianshan)
N	Nanshan Bei Lu [Nan-Shan Pei Lu]	南山北路	(Road north of the Nanshan)
P	Liuzhong Lu [Liu-chung Lu]	柳中路	(Road through the Willows)
R	Lingshan Dao [Ling-shan Tao]	霊山道	(Magic Mountains Road)

the really heroic test of the ancient road-builders was the challenge of the Qinling Shan, that great range of mountains shutting off Chang'an (1) and the Wei valley from the south-west. By the end of the century, the Qin rulers were pushing on their colonisation across the barbarian uplands of Yunnan. Sima Qian, the Han 'Historiographer Royal', says laconically: 'Under the Qin, Chang An planned and constructed a Five-Foot Way.' Starting across the Yangtze from Ba (84, modern Chongqing), it ran down the hills of Guizhou via Yelang (102) and Langzhou (103) to reach the Kunming Lakes at Dian (90) and Yizhou (104) and then on to the Erhai Lake at Yeyu (91). This was a traverse very similar to that of the present-day road between Kunming and Chongqing, and continuous with the Burma Road, which took so much traffic during the Second World War.

If the Qin capital had five major imperial roads radiating from it, the same metropolis under the Former Han emperors, rulers of a still more closely knit sub-continental nation, had as many as seven. To the north-west, there was a further opening up of the Gansu and Shaanxi hill-country north of the

Wei valley, with new roads to Anding (50) and Huizhong Gong (53), the latter being completed in 108 BC. To the south-east, the distance to Changsha (16) and points south was shortened by a road direct from the capital to Xiangyang (71). Northward, the capital was connected to Jiuyuan (5) by a road, which, though more circuitous, passed through easier and more inhabited country than Meng Tian's Great North Road, which now lay on its west side. However, of greater importance than any of these was the use of the long Fen River valley in Shanxi to throw a road north-eastwards to Beijing (7); considerably shorter than the previous one, it still necessitated crossing the Yangtze.

In the eastern region, the Han road-builders were particularly active, laying down a network of highways all over the North China Plain. Further extensions still were made by the Later Han, though in ancient times no penetration was made of the Fujian amphitheatre, where the country was too difficult and the people still barbarous.

Out towards the west the case was very different. The desert oases formed less of a barrier to intercourse than the forested mountains, and the moment has come to glance at the Chinese end of the Old Silk Road, an incurably romantic subject. We must see how the Chinese highway network joined the caravan trails around the Taklamakan Desert in Xinjiang (Chinese Turkestan), starting where the complex of ancient roads ends in the Wei valley at Baoji (23). From here, the road passes through the gorges of the upper waters of the Wei at Tianshui (24), then up over mountain passes and down to Jincheng (27, modern Lanzhou) on the Yellow River beyond the great bend. For two millennia, this was the hub of routes of the western region.

The road then goes over another pass at Wushao Ling (29) to Wuwei (30), anciently known, in error for Chang'an, as Sera Metropolis. Next comes Yongchang (31), site of the settlement in 30 BC of captured Roman legionaries, and, after crossing many streams and alluvial fans, Jiuquan (35, modern Suzhou).

So far, the road has been protected on its right-hand side by the extension of the Great Wall, also constructed during the Han period. At the same time, a road was made north-eastwards to the lost city of Juyan (99), certainly another listening-post for tribal intentions. Now the road leaves the guardianship of the Wall and strikes off across wild desert country to Anxi (38, anciently Guazhou, the Melon City), a fortified post marking the first fork of the Old Silk Road, this whole 'Gansu Corridor' being settled after the victory over the Huns in 121 BC.

Thenceforward, we have to consider circumvention of the Taklamakan Desert in the Tarim Basin by two main routes. To the south, there was the Road North of the Southern Mountains (Nanshan Bei Lu, the enormous Tibetan Gunlun range). This forks at Dunhuang (39, anciently Shazhou, the

City of the Sands), one (N) keeping to the foothills to reach Khoti, Kashgar and the West by way of Yuni (43), the other passing through the lost city of Loulan (42) beside Lake Lop Nor. West of Dunhuang and Yumen Guan (The Jade Gate), the routes were caravan trails rather than roads. The northern route round the Taklamakan Desert lay south of the Celestial Mountains (Tianshan Nan Lu) and ran along their foothills (M) to Kashgar via Yanqi (47), though towards the end of the Former Han there was another route from Anxi (38); this (P) was known as the 'Road through the Willows'.

The greatest engineering feat of the Qin and Han road-builders was the consolidation of the passes through the towering Qinling mountains, along those routes pioneered by the Qin people as early as the fourth century BC. Their problem had been to find ways across the great mountain divide between Shaanxi in the north, where the Qin capital was located, and the Sichuanese basin to the south, where the half-barbarian States of Shu and Ba, occupying rich agricultural land in a key economic area, were inviting annexation from the aggressive Qin State. A deep intervening valley, that of the Han River, allowed the surveyors to take their breath before finding their way down or across the head-waters of the Jialing River and into Sichuan. First, then, there was the range which brings Taibai Mountain with its snows into sight from every vantage-point on the northern slopes of the Wei valley, and afterwards there were the less grim ranges of Micang and Daba. Even now, the motor road which connects Sichuan with Shanxi passes through remarkable scenery, but in ancient times it must have been even wilder and more difficult country.

The focal way-stations were three small cities on the upper reaches of the Han River: Mianxian (72), Baocheng (73) and Hanzhong (74). All the roads which went through them were in existence when the First Qin Emperor ascended the throne in 221 BC, but they needed, and received, great improvement under the Han. The oldest road (F) had started from Chencang (23), crossed the Wei river, ran up a small tributary to the little town of Fengxian (75), then turned west to the upper waters of the Jialing until it doubled back into the mountains to find its way to Mianxian (72). The first Qin improvement gave route B (the Linked-Cloud Road); this was to shorten the western detour by striking through the mountains directly from Fengxian (75) by way of Liuba (76) to Baocheng (73). The passes on this route did not exceed much more than 2000 metres, but no less than one-third of its 215 kilometre length was composed of wooden trestles built over the beds of roaring torrents, or actually shelved into the cliff higher up by means of timber brackets driven into holes in the rock face. A Western parallel was the mountain road building of Tiberius (*ruled* AD 14 to 37) and Trajan (*ruled* AD 98 to 117), though these feats were rather limited in comparison. It is interesting that even now the modern motor road follows approximately the route of the Linked-Cloud Road, which has been the principal means of communication through the mountains since the Song dynasty.

The second Qin improvement (C) chose a still more easterly traverse. Starting from Meixian (77) on a road running south, not north, of the Wei river from Chang'an (1), and then travelling up the steep valley of the Ye tributary through the Yeyu Pass, it came out on the high country surrounding Taibai Mountain to the west. Here it found the southward pointing valley of the Bao or Taibai River, where a place still bears the name Tong-che-ba (Passing-place for Carts), and dropped down to Baocheng (73) from near Liuba (76). It thus short-circuited the Lianyun Road (B). First constructed with innumerable trestle and gallery sections in the fourth century BC, it was extensively enlarged and repaired in about 260 BC, then again in 120 BC and AD 66. Because of its engineering works, it became known as the Beizhan Lu (the Northern Trestle Road). Early in the third century AD, Zhuge Liang described the massive pillars and beams that upheld the road through the ravines which it traversed, but about what happened right back in 120 BC we know a good deal because Sima Qian says:

> After this, someone presented a memorial to the Emperor proposing that the Baoye Road should be adapted for the transport of grain by boat. The matter was referred to the Censor-in-chief Zhang Tang, who enquired into the matter and reported as follows: 'In order to reach Shu (Sichuan) (traffic) goes through the Marches of the Old Road (Gudao) where there are many rocky descents and long detours; if now we pierce a (better) road between the Bao and the Ye the gradients will be much less difficult and the distance will be shorter by 200 kilometres. Since the Bao river flows into the Mian river, and since the Ye descends into the Wei, we should be able to use them for grain transport by boat. The grain would come up from Nanyang along the Han, the Mien and the Bao; then from the point where the Bao becomes too shallow the grain will be transported in carts to the headwaters of the Ye about 50 kilometres or more, after which other boats will take it down the Wei (to the capital). Thus the grain of Hanzhong will become available, and that from the east of the mountains can come in unlimited quantities, much more easily than it does past Dizhu. Lastly, the abundance of wood and bamboo, both small and large, in the Bao and Ye valleys, useful for construction, is comparable to that in Ba and Shu itself.' The emperor approved the project.
>
> Zhang Tang's son, Zhang An, was (accordingly) appointed Administrator of Hanzhong, and he recruited several tens of

thousands of men who reconstructed the Baoye Road over a total distance of more than 250 kilometres. The road was in effect convenient and shorter (than the others), but the rivers proved too violent and too much encumbered with rocks to allow of the grain transport by boats which had been envisaged.

This passage is of much interest, as it illustrates the close coordination between land and water transport which characterised Chinese planning for more than two millennia.

Between the rules of Tiberius and Trajan, the Baoye Road, incorporating the southern portion of the Lianyun Road, was repaired again after some three centuries of continuous use, and what that involved may be estimated by the figures which have come down to us – 766 800 man-days of work, carried out by 2690 convict workers. In Han times there were also *corvée* labourers, and, as well as the kilometres of trestles and galleries, the road possessed 623 small bridge-spans and five large ones, while sixty-four post-station rest-houses were provided along a stretch of 129 kilometres.

Besides the roads already described, there were two more easterly ways over the Qinling mountains. One of these (D) started from Zhouzhi (78), followed the Liuye River, a tributary, then came out in high country, east, not west, of Tai-Bai Mountain. After this it found its way to Hanzhong (74) over passes ranging from 2000 to over 2700 metres high, through the Luo Valley. But in later ages it was little used, except perhaps during the Tang period.

A still more easterly route (E) was taken by the North–South Road, which went straight up into the hills south of the capital, and, passing over a high plateau, made for the valley of the Ziwu River, whence Hanzhong was reached on either bank of the Han River. This route was pioneered in the time of Wang Mang (AD 5), but as late as the eighth century AD it was still a regular courier route.

Not only in the Qinling mountains, but all over China trestle work and galleries were a prime feature of road engineering through the centuries, so mountainous is the country. In the seventeenth century AD, the Jesuit missionary Louis Lecomte could still write:

> The Civil Government of the Chinese does not only preside over the Towns, but extends also over the Highways . . .
>
> The Road from Chang'an to Hamtchoun is one of the strangest pieces of work in the world. They say, for I myself have never seen it, that upon the side of some Mountains which are perpendicular and have no shelving they have fixed large beams into them, upon the which beams they have made a sort of Balcony without rails, which reaches thro' several mountains in that fashion; those who

Fig. 391. A stretch of one of the trackers' paths or half-tunnel towpaths cut in the rock faces of the Yangtze gorges, here at the defile known as the Wind-box Gorge, above Yichang. (Photo. Popper.)

are not used to these sort of Galeries, travel over them in a great deal of pain, afraid of some ill accident or other. But the People of the place are very hazardous; they have mules used to these sort of Roads, which travel with as little fear or concern over these hideous precipices as they would do in the best of plainest Heath.

A thousand years earlier, the great poet Li Bai, who passed over the Qinling Mountain roads more than once, wrote a famous poem on 'The Road to Sichuan', in which he said that '. . . It would be easier to climb to Heaven than walk the Sichuan Road'.

The construction of cliff-galleries was also naturally adopted in Tibet. We hear of 'ancient suspended roads' in the records of the travels of the Korean Buddhist monk Huizhao to and from India in around AD 726 on his pious missions. The Chinese also used half-tunnelling, i.e. excavation of half the cross-section of the road into the cliff-face with a rock overhang, just as the Romans did during the reigns of Tiberius and Trajan. Sometimes the rock will support an almost full cross-section, and astonishing examples of great length are seen in the paths of the trackers or haulers of boats through some of the gorges of the Yangtze (Fig. 391), while modern motor roads have also made much use of the technique. It is probable that the balcony roads of the Chinese were sometimes suspended from chains, since, as we shall see, the invention of the iron-chain suspension bridge occurred so early among them (see pp. 151 ff.).

It would be hard to overestimate the importance of the Qinling passes in Chinese history. Twice at least these mountain roads became the paths of emperors seeking the sanctuary of mountain-battlemented Sichuan. But much more important was that, after the assimilation of Ba and Shu by the Qin in the fourth century BC, that State had become possessors of the vast natural resources of Sichuan. This, together with great irrigation projects which the Qin also completed, must have been a cardinal factor in the first political unification of all China. And, later, the Old Silk Road was fed by the Linked-Cloud trestle ways, for much of the silk exported to Rome was Sichuanese in origin.

In the second century BC, much wider horizons began to open. The emperor Han Wu Di began to consider annexing the Cantonese State of Nan Yue and the other smaller States of the south from Fujian to Annam (now part of Vietnam). His ambassador at Canton in 135 BC, Tang-Meng, noticed that produce of Sichuan found its way down the West River, and argued that, in that case, naval and military forces could come too. But to do this, a road would have to be constructed to the Zangke River (92). Accordingly, in 130 BC building such a road was undertaken by several thousand conscripts, for whom provisions and meals were carried on the backs of porters for 500 kilometres, though only about 2 per cent reached their destination. Long and hard in the building, the road met with the opposition of local princes, who at first prevented its use; but, less than twenty years later, boats of more than one deck, or their parts, were being carried along it to help the forces fighting the southerners. Yet though this new southern road was intended to transport men and supplies southwards from Podao (89) in Sichuan through Yelang (102) in Guizhou to Guangxi and Guangdong, it had the effect of opening up Yunnan also.

Now it only remains to sketch a few of the major developments of the Later Han period. In AD 27, an important strategic road (J), over 150 kilometres long, on the northern frontier, was built by Du Mao, a military engineer who also constructed many fortifications, beacon stations and transport depots. Starting from the old city of Dai (61), it led up through the Feihu Pass, from which it gained its name, 'The Flying Fox Road', over the uplands north of Mount Wutai, so famous for its mountain abbeys, to reach Pingcheng (86) near Datong, a focal point for the guard of the Great Wall. This was the region in which another noted engineer, Yu Xu, stationed between Lueyang and Zhaohua, found that:

> the roads carrying traffic in (that district) were very difficult and dangerous. Neither boats nor carts could get through (the narrow passes), so donkeys and pack-horses were used, which cost five times the value of all that was transported.

Therefore Yu Xu himself led out his officers and clerks on tours of inspection through the river gorges from Chu down to Xiabian, and he caused the rocks to be cut down for several tens of *li*, so as to open the road for boats to pass.

He thus constructed what was probably a tracker's path, able also to take some wheeled traffic, beside a river which had been made navigable. A commentator explains:

> To the east of Xiaben for more than 30 *li* there is a gorge, where the rocks formed great barriers in the way of the spring overflows. This led to floods in spring and summer, spoiling the harvest and damaging the villages. So Yu Xu made his men set (great) fires to the rocks and then lead water on to them, so that they split in pieces and could be removed with crowbars. Afterwards there was no more worry about flooding.

This is far from being the only mention of the fire-setting method in Han texts. Indeed, the resolution of the old highway engineers, ill-equipped as they were, can be guessed from the description of Li Xi directing the burning and splitting of rocks in about AD 160.

But perhaps the most extensive Late Han road building took place in the south. In AD 35, a road was thrown along the mountains north of the Yangtze from Ba (84, Chongqing) to Zigui (88) in order to bypass the dangerous Wu mountain gorges. There was also a very much larger network which took the form of a cross in Jiangxi, Hunan, Guangdong and Guangxi. The northwest to south-east limb started from Guiyang (98) and went via Nanhai (70, Canton) to Yingde (97). The north-east to south-west limb ran from Qingjiang (67) to Gugong (69), much as a good motor highway does now. This system was complete by AD 31. Subsequently, developments were more military in conception, and by AD 83 the second limb was extended all the way to Hanoi in North Vietnam. Meanwhile, in AD 41, General Ma Yuan built a mixed land-and-water route with a road 500 kilometres long running from Hepu (94) in the Leizhou peninsula north of Hainan island, westwards to Jiaozhou (95).

We are now in a position to make a very rough assessment of the extent of the road building in the Qin and Han periods. The approximate distances amount to just under 65 000 *li* (32 500 kilometres). However, the circuitousness of some routes must have been considerable, and we shall not be wide of the mark if we accept a final estimate of between some 32 000 and 42 000 kilometres for the main roads in existence by the end of the Han. Estimates of the extent of the Roman roads vary considerably, but seem to indicate something of the order of 78 000 kilometres, with nearly 4000 kilometres in Britain and almost 14 500 kilometres in Italy. Taking into

Table 53. *Road building in the Chinese and Roman Empires.*

	Area (square kilometres)	Road length (kilometres)	Road length (kilometres) per 1000 square kilometres
Roman Empire			
Trajanic (before AD 117)	5 084 000	78 000	15.3
Hadrianic (after AD 117)	4 566 000	78 000	17.1
Gibbon's estimate	4 144 000	78 000	18.8
Chinese Empire			
Late Han (c. AD 190)	3 968 000	35 400	8.9

account the changing area of the Roman Empire over the years, Table 53 shows that the two road systems are quite comparable. But it may be that the Roman figures are too high, for, especially in North Africa and much of the Near East, the Roman paved military roads probably tailed off into caravan tracks like those of Xinjiang and other 'Western regions' not included in the Chinese figures.

Broadly speaking, however, it may be said that, down to the third century AD, the Chinese network attained from 55 to 75 per cent of the Roman one. In this difference, both geographical and political circumstances may have been important. The lesser relative mileage of the Chinese network must surely have had something to do with the greater navigability of the inland rivers and the greater use of artificial waterways in China when compared with the situation in Europe. The Roman road system was, in a way, an external skeleton, for the heart of the Empire was a vast (and potentially stormy) body of water, the Mediterranean Sea. Though this certainly facilitated maritime transport, the Romans seem still to have felt the necessity of laying down roads of great length all round the basin. On the other hand, the Chinese highway system radiated out from a particular centre, Chang'an (1), over a very large continuous land mass, constituting, as it were, the internal skeleton of the Empire. In any case, we have every reason to admire the planning and construction of the ancient Chinese roads, the firmness of which has not entirely yielded to the effort of fifteen centuries.

There is space for only a very brief reference to the subsequent development of the Chinese highway system. After its rapid rise during five formative centuries, it declined in importance relative to the waterways during the following ten. The last mention of government relay carriages occurs in AD 186, and for official travel they were replaced by the post-horse system, carts being used only for the transport of goods and for the movement of families of the better sort. At the same time, the Later Han suffered a shortage of horses, Luoyang having but one stable instead of the six which the capital of Chang'an had possessed two centuries earlier. Wagons drawn by oxen, mules or donkeys came into more widespread use, and on the feeder roads and pathways the South Chinese custom of riding in litters borne on men's shoulders prevailed more widely. Indeed, a connection of this practice with the invention of the wheelbarrow in the Late Han (this abridgement, volume 4) has been suggested.

The pattern of government enterprise nevertheless persisted. Certainly, roads were not inexpensive, and, during the Early Han, figures are similar to those for Roman roads. But, as the centuries went by, there was a general tendency to replace civil by military control, though use was made also of convicts and unpaid labour in place of taxes (*corvée* labour).

Though the network decayed, it was not despised even during the period of disunion of the Northern and Southern Empires (*c.* AD 380 to 580). There were notable road-builders in the fifth century; for instance An Nan, who built roads for the Northern Wei dynasty, while in AD 493 a Regius Professor of Geographical Communications was established. Then, early in the seventh century, during the Sui dynasty, Yang Di constructed a military road some 1500 kilometres long and 30 metres wide running parallel with the Great Wall, which he extensively repaired all the way from Yulin (56) to Beijing (7).

The road network during the Tang (AD 618 to 907) was very thoroughly organised, yet its total length was only some 218 000 kilometres, a distinct reduction from the Late Han figure. It also showed significant differences with the Qin and Han routes of Fig. 389. Apart from a new route into Yunnan running south from Yuesui (93), that province, together with those of Guizhou and Guangxi, had almost reverted to nature. On the other hand, the upper and lower Yangtze Valley regions were connected by a road bypassing the gorges completely, while the network between the provinces Jiangxi, Hunan and Guangdong was well kept up. We also see, for the first time, a highway running down from Wu (11) in Zhejiang to the great ports of the Fujian coast, which were to be such famous centres of international trade during the Song. North of the 23rd parallel, the Old Silk Road, the Qinling passes, the Road of the Golden Oxen, the northern highways to the Wall and to Beijing, as well as the ways to Nanyang (14) and Shandong province, all radiating from the eternal city Chang'an, remained very much as they had been in the youth of the Chinese Empire.

Military interests were not the only factor which interested government circles in roads, and indeed waterways as well, for some of these played a significant part in the transportation of tax-grain and in carrying official messages. However, the upkeep of a multitude of local roads and paved pathways devolved upon the people themselves, acting in their cooperative capacity under village elders and small-town worthies. In this context, religious associations, such as the Daoist Yellow Turbans in about AD 180, later politically so important, or the Buddhist fraternities afterwards, played an important part. Indeed, there was a long tradition of such privately initiated roads going back to the Han or even earlier, and the total length of these roads far outstripped that of the government main roads as the ages passed.

Foreigners were usually much impressed by China's land communication system. The Japanese monk Ennin, who travelled widely during the Tang between AD 838 and 847, had many things to complain of, but never deprecated the roads, with their milestones, signposts, watch-towers, ferries and bridges. Though nineteenth-century travellers grumbled about the Chinese highways, which fell into a decline under the Qing while those of Europe steadily improved, the roads inspired the admiration of Westerners in the seventeenth century. After the gloomy accounts of merchants and missionaries, it is rather surprising to read Lecomte's words of 1696:

> One can't imagine what care they (the Chinese) take to make the common Roads convenient for passage. They are fourscore foot broad or very near it; the Soil of them is light, and soon dry when it has left off raining. In some Provinces there are on the right and left hand Causeways for the foot Passengers, which are on both sides supported by long rows of Trees, and ofttimes terrassed with a wall of eight or ten foot high on each side, to keep Passengers out of the fields. Nevertheless these Walls have breaks, where Roads cross one the other, and they all terminate at some great Town.

The post-station system

Once the work of the road-engineers was completed, it remained to weld it into a great social institution by establishing and deploying a whole army of messengers, coachmen and station-masters along the lines of communication throughout the Empire's length and breadth. Such a system arises naturally as soon as a developing society reaches an imperial level of organisation. It appeared in fifth century BC Iran, descended to Egypt of the Ptolemies, and reached its western peak in 31 BC under Augustus.

The Chinese possessed a suitable basic system for such a development, and Shang records of the fourteenth century BC – comparable to those of Babylonia and Ancient Egypt – speak of systematic reports from the frontier regions. This gives colour to the definition of the term *yu* in the Late Han *Shuo Wen Jie Zi* (Analytical Dictionary of Characters) as 'a station on the border for the transmission of despatches'. By the time of the compilation of the *Zhou Li* (Record of the Institutions of (the) Zhou (Dynasty)) in the second century BC, the system is taking the very definite shape which it conserved for 2000 years. Referring to the Almoners-General (*Yi Ren*), the text says:

> In principle, along all the roads of the Empire and the (feudal) States there is a rest-house every 10 *li* [some 5 kilometres] where food and drink may be had. Every 30 *li* there is an overnight rest-house with lodgings and a government grain store. Every 50 *li* there is a market and a station with an abundant stock of supplies.

Such stations each had a watch-tower, while at the overnight rest-houses there were stables besides human lodgings.

Broadly speaking, the main roads were equipped from Han to Song times with a post-office every 5 *li*, a cantonal office every 10 *li* and a post-station every 30 *li*. These short distances were undoubtedly chosen so that flag and drum signals, or the fire and smoke of beacons, could readily be seen. The postal clerks kept records of the despatches they transmitted – in Han times they were written on 30 centimetre-long wooden strips contained in bamboo tubes closed with a spring lock contrivance – and the cantonal officers policed the road and its neighbouring districts with guards. At the post-stations, there were stables and couriers in readiness for the relay service, under the authority of a station-master. Strategic points in the network had veritable sorting-offices, where mail was collected and distributed. By 77 BC, the post-station system extended as far west as Loulan (42) and Baoshan (100). By AD 89, tropical fruits were sent with the aid of the relay service from Nanhai (70) to the capital. The post-station rest-houses had washing and sleeping accommodation for officials and authorised travellers, as well as restaurants, and also cells for prisoners moving under guard. Often, the way-stations had private inns or hostelries for those who lacked the right to use government facilities.

The post system reached a particularly high development in the Tang period, with a force of some 21 500 officers strung out along the roads and managed by 100 high officials. In the Song, however, the post-station service had become the 'hot-foot relay' (*ji jiao di* [*chi chiao ti*]) and the stations 'horse relay' points (*ma di* [*ma ti*]) or 'horse stables' (*ma pu* [*ma phu*]). With the Yuan dynasty, everything became increasingly militarised, the couriers being under the orders of controllers. Though corruption was prevalent at all levels, the service in the fourteenth century AD actually attempted to work a timetable. After this, there was little change until the arrival of the telegraph and modern road building in the nineteenth century.

The speeds attained by the couriers appear to have averaged some 190 kilometres in 24 hours, that is about 8 kilometres per hour, somewhat akin to that of the Roman post-couriers. Rather more could be obtained by men who galloped 12 hours from dawn to dusk at an average speed of almost 18 kilometres per hour. However, the literature contains a number of exceptional records for the Liao and Qing periods, ranging to over 400 kilometres a day, explicable only by conditions which permitted continuous travel throughout the night. Beacon signalling alone gave faster results. Thus, in 74 BC, news of the death of the emperor, Zhao Di, was transmitted by this means at a speed of some 43 kilometres per hour.

The bureaucracy was always sensitive about the misuse of transport, and one of the crimes with which General Yang Pu was charged in 111 BC was that of asking for a relay carriage to go to the frontier, but actually using it to go home instead. Public transport was correspondingly a mark of imperial favour, as, for instance, with a scientific–technological conference in AD 5 when government coaches were provided for those physicians attending. This followed a precedent established for delegates about eight years earlier. We know little for certain of the speed attained by the relay carriage system, but to judge by records of an urgent journey made by the Prince of Changyi in 74 BC, speeds as high as 14 kilometres per hour were reached. Horses were changed every 30 *li* at the post-stations.

Administratively, a number of departments were involved in operating the system. During the Han, carriages and mounts were under a Grand Keeper of Equipages, but another bureau, the Regulation Department, looked after the post system, while the Commandant's Department was in charge of public vehicles. Then, each province or commandery had its Superintendent of Posts and was divided into a number of postal districts. In addition, each provincial government had civilian Secretarial Aides for Highways and Bridges and Fords Officers. There were also military Commandants of Passes and Barriers, who were responsible for checking the credentials of travellers and goods, collecting internal custom fees, preventing smuggling and maintaining general security. All passports required the signature of the Prefect of Credentials and Tokens. The *corvée* and taxation system always applied for the needs of the post service, local people having to pay special taxes for the horses, and to supply a tithe of food for the travellers. But the people disliked *corvée* and direct taxation, and adopted a passive hostility to the road authorities and all their works. Furthermore, the tribal peoples saw very clearly that the road network was a major factor in the expansion of the Chinese Empire, and therefore cut the roads and destroyed the post-stations whenever they could. Yet, in spite of everything, the Chinese highway network, and the post-station system which was inseparable from it, constituted a cardinal factor in the advance of East Asian civilisation.

2

Walls, and the Great Wall

There can be no doubt that the most ancient form of walling in China, both for houses and un-roofed enclosures, was that of *terre pisé*, or tamped earth. Here, elongated boxes without tops or bottoms were used, and into them dry earth was rammed at successively higher levels as the wall rose (Figs. 392, 393). This 'shuttering' resembled that used today for confining concrete while setting, and the whole tamping process was so basic in Chinese building technology that the word for it, *zhu [chu]* (築), came, in course of time, to form one-half of the standard term for architecture in general (*jian zhu [chien chu]* 建築). This can be seen to be appropriate when one remembers that the platform foundations of ancient buildings were also of rammed earth.

In Chinese practice, it was customary to use rubble stone without binding material as the foundation of walls, and to spread a layer of thin bamboo stems between each *pisé* block so as to hasten thorough drying-out. The prevalence of *terre pisé* in Chinese culture certainly has something to do with two features which we shall later find characteristic of Chinese architecture: first that the walls were not, in general, weight-bearing; and secondly that buildings were provided with generously overhanging eaves.

In modern times, travellers have seen so much more *terre pisé* walling than in their own continent that they have thought of it as a Chinese invention. However, the process was well known to the Roman, Pliny the Elder (AD 23 to 79). Indeed, in rural England and France, these techniques have never died out, and in many English counties boundary walls with roofs of tiles or thatch, looking extremely Chinese, can be found.

In late Zhou times, bricks were mostly of adobe, i.e. sun-dried mud, such as may frequently be seen in China still, but, by the Han period, baked bricks were becoming general. Plaster was then already in use, often covered with paintings. Many different sizes of bricks have been current throughout the ages, ranging from 30 centimetres × 23 centimetres × 15 centimetres in the north-east, through to tile-like shapes, which in the Great Wall measure

Fig. 392. Tamped earth (*terre pisé*) walling under construction; a drawing from the *Er Ya* (Literary Expositor).

Fig. 393. Tamped earth (*terre pisé*) walling under construction, with poles used as shuttering, near Xi'an. (Photo. Joseph Needham, 1964.)

Fig. 394. Box bond brickwork in the walls of the old Luo family temple beside the Chongqing–Chengdu road, Sichuan. (Photo. Joseph Needham, 1943.)

38 centimetres × 19 centimetres × 3.8 centimetres. As far as laying them is concerned, besides the ways the different layers overlap – the 'bonding' – with which we are familiar in the West, the Chinese have long used a 'box bond'. Here, stretchers are placed vertically between layers of horizontal ones, the interior being filled with earth and rubble (Fig. 394). Sometimes two horizontal layers intervene (Fig. 395). There is even a Chinese cross bond in which the stretchers are placed in groups of three.

As well as the ordinary baked bricks of the Warring States and Han periods, the Chinese craftsmen were the first to master the art of moulding and firing blocks of terracotta brick, ornamented with very intricate scenes and patterns (Fig. 396). These were used mainly for walling tombs, and those in Han tombs afford many glimpses of life at that time. Yet at a later time they were used in fortifications. Indeed, it may be worth quoting from a curious passage concerning the use of baked bricks for fortification at the capital of a short-lived barbarian dynasty, the Xia (AD 407 to 431), whose Hunnish rulers governed parts of Gansu and Shanxi. The *Jin Shu* (History of the Jin Dynasty) says:

> In this year (AD 412) the reign name was changed to Fengxiang.
> (The emperor) Helian Popo chose Chigan Ali to take office as
> Chief Engineer, and to mobilise 100 000 workers of the Yi and Xia

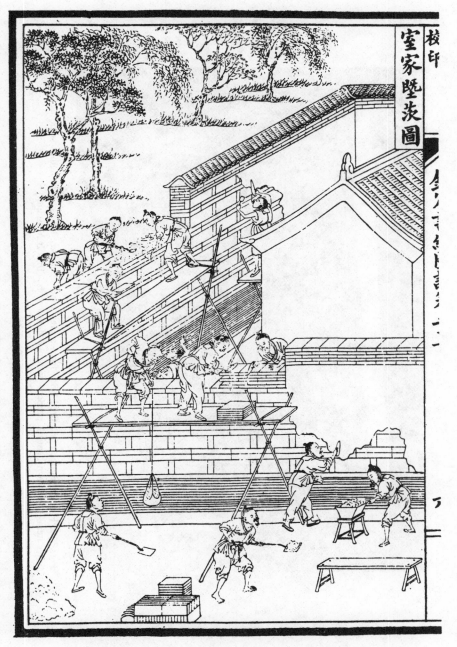

Fig. 395. A late Qing representation of bricklayers at work. Box bond with horizontal stretchers in double layers and a coping of five layers of headers may be seen. Plasterers are also present. From the *Qinding Shu Jing* (The *Historical Classic* with Illustrations) of AD 1905.

Fig. 396. Hollow stamped and fired brick from a Han tomb-chamber. Two human figures standing and other items are shown beneath porticos with two-tiered roofs. (British Museum.)

peoples north of the mountains. They were to go to a place north
of the Shuofang River and south of the Heishui River, there to
build the capital. Popo himself said, 'As I have just united the
world and attained the empire over ten thousand regions, I shall
name it Tong-Wan.'

   (Chigan) Ali was extremely skilled and clever, but also cruel and
violent. He caused the workers to bake bricks to make the city
wall. (He used to test the bricks) and if a hammer blow would
make a depression as much as 7 centimetres deep, he would have
the worker (responsible) killed and buried inside the wall. Popo
thought that this showed much loyalty on the part of Ali, and gave
him the responsibility of all buildings and repairs.

Chinese literature contains, or did contain, an illustrated treatise especially devoted to the manufacture of bricks and tiles. This was the *Zao
Zhuan Tu Shuo* (Illustrated Account of Brick-and-Tile Making) of Zhang
Wenzhi, written at some time during the Ming dynasty in the second quarter
of the sixteenth century AD. Zhang was an official of the Ministry of Works
who had responsibility for certain State brickfields, and he wrote his book
partly in order to remedy the bad conditions prevailing among the workers
and partly to avert the confusion of organisation, which had led to the
suicide of some of the independent contractors. But the Ming dynasty decayed before much of his reform could be achieved.

The importance of walls in ancient times is shown by the fact that several
words were used to describe different forms of them. High walls round
courtyards were called *qiang* [*chhiang*] (牆) or *yong* [*yung*] (墉), house
walls and party walls were known as *bi* [*pi*] (壁) and low walls in gardens
etc. were termed *yuan* (垣). In the 1920s, the architectural historian O.
Sirén gave a graphic description of the walls of traditional China:

   Walls, walls, and yet again walls, form the framework of every
   Chinese city. They surround it, they divide it into lots and
   compounds, they mark more than any other structures the basic
   features of the Chinese communities. There is no real city in
   China without a surrounding wall, a condition which indeed is
   expressed by the fact that the Chinese used the same word *cheng*
   [*chhêng*] (城) for a city and a city wall; there is no such thing as a
   city without a wall. It would be just as inconceivable as a house
   without a roof. These walls belong not only to the provincial
   capitals or other large cities, but to every community, even small
   towns and villages. There is hardly a village of any age or size in
   northern China which has not at least a mud wall, or remnants of
   a wall, around its huts and stables. No matter how poor and

> inconspicuous the place, however miserable the mud houses, however useless the ruined temples, however dirty and ditch-like the sunken roads, the walls are still there, and as a rule kept in better condition than any other constructions. Many a city in north-western China which has been partly demolished by war and famine and fire, where no house is left standing and where no human being lives, still retains its crenellated walls with their gates and watchtowers. Those bare brick walls, sometimes rising over a moat, or again simply from the open level ground where the view of the distance is unblocked by buildings, often tell more of the ancient greatness of the city than the houses or temples. Even when such city walls are not of a very early date . . . they are nevertheless ancient-looking, with their battered brickwork and broken battlements. Repairs or rebuilding have done little to refashion them or change their proportions.

Though Sirén wrote this several decades ago, and the exigencies of modern life have now led to the disappearance of some parts of city and village walls, it will be a long time before travellers cease to marvel at their size and ubiquity.

Even the ancient Asian nomads surrounded their camps with earthen ramparts, and there can be no doubt that the essentially settled agrarian culture of China erected walls round its earliest cities. Indeed, the character for capital, *jing*, [*ching*] (京) has as its oldest graph (髙) a picture of a guard-house over a city gate. And city walls are to be found as early as the fifteenth century BC. Those of the Shang city of this period, Ao, just north of modern Zhengzhou, are some 20 metres wide at the base and enclose an area of about 1920 metres square. Similar excavations and studies have been made of Zhou feudal capitals, such as that of the State of Zhao at Handan in Hubei province, founded in 386 BC, where the main rectangular enclosure has sides some 1399 metres in length, with walls originally as high as 15 metres on a base 20 metres wide. All such walls consisted essentially of successive layers of tamped earth (*terre pisé*) averaging from about 7.5 to 10 centimetres thick. Indeed, when Chinese city walls are cut through today for modern transport improvements, these layers can still be seen, though whether or not the walls of Shang and Zhou cities were always faced with sun-dried (adobe) bricks we cannot be sure.

By the end of the third century BC, the art of fortification had made much progress, and the unification of the empire provided such an abundance of men and materials that the walls of the Western Han capital at Chang'an, still traceable some 8 kilometres north-west of modern Xi'an, were on an altogether greater scale. Along a circuit of some 28 kilometres, arose a

rampart wall 15 metres high and 12 metres wide at the top, without fortifications; but it did not stand alone. It was backed by a flat area, or terreplein,
more than 60 metres wide, raised about 6 metres above the surrounding
land, and protected by a moat about 46 metres wide and 5 metres deep.
This terreplein was once covered with buildings, perhaps garrison dwellings,
and part of this earlier structure can still be made out. The distance from
the outer edge of the moat to the foot of the inner face of this terreplein
must have been of the order of 140 metres. All this was an effective reply
to the techniques of besieging a city which had developed during the Warring
States period and included fire-setting in tunnels to cause the collapse of
walls, or the diversion of rivers to wash away their foundations. The traditional builder of these fortifications in about 200 BC was Yangcheng Yan.

The cores of Chinese city walls were always of earth or rubble, but in later
centuries they were usually provided with outer, and also often inner, facings
made of large grey burnt bricks laid in lime mortar. Occasionally, where
stone was plentiful, as in Sichuan, walls were faced with dressed stone
blocks in regular courses of equal height. A traditional Chinese representation of the foundation and facing of a city wall is shown in Fig. 397, where
we see also the in-fill of rubble. Part of the walls of Xi'an are illustrated in
Fig. 398.

Chinese city walls were never complete without their watch-towers and
gate-towers, usually single structures in two or three storeys, with the up-
turning roof corners so characteristic of Chinese buildings, and set directly
over the opening in the city wall. Such openings are sometimes enclosed
by a curving wall standing on its own (a curtain wall). An alternative type
of city entrance, probably rather older, flanked the gateway by two such
pavilion-bearing towers. But whatever may be the plan, walls and bastions
invariably slope markedly inwards to the top, in contrast to the perpendicu-
lar walls so often seen in medieval Western castles. Fig. 399, a Tang period
fresco from Dunhuang, shows such typical Chinese defensive architecture.

The massive Chinese wall did not always rise straight out of the ground
or the water-filled moat. It frequently had a supporting platform or plinth,
just as if it were any other building (see p. 49). But since many of the plinth
designs had indented mouldings, placing a heavy wall above them gave a
remarkable effect of elegance and lightness. As a result, the spreading static
effect of Western classical designs is altogether avoided.

The Great Wall

The Great Wall (the Ten-Thousand *Li* City-Wall – *Wan Li Chang Cheng* as
it is called in Chinese) has probably been in the reader's mind during the
previous paragraphs. It notably stirred the imagination of eighteenth-century
Europeans. In AD 1778, Dr Johnson

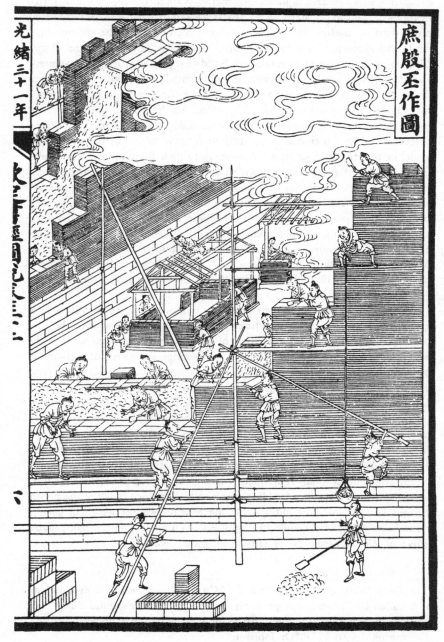

光緒三十一年

庶殷丕作圖

Fig. 397. A late Qing representation of the construction of the walls of the capital city of the Zhou dynasty. On a solid foundation large bricks or dressed stone are the rampart walls of tamped earth; these are contained by headers and stretchers of smaller thin bricks which are rising into the clouds. From *The Historical Classic with Illustrations* of AD 1905.

Fig. 398. The walls of Xi'an in 1938. In the foreground is one of the ramps giving access to the ramparts. As in many other cities, shrinkage of population had made room for much agricultural land within the walls. (Photo. from Bishop (1938).)

Fig. 399. The style of Chinese defensive architecture seen in a Tang fresco at Qianfodong. Notice the inward sloping wall surfaces and the galleried barracks pavilions surmounting corner towers and gate-towers. The dresses and uniforms of the period (*c.* AD 660) are also noteworthy. (Photo. Lu Jimei, 1943.)

... talked with an uncommon animation of travelling into distant
countries; that the mind was enlarged by it, and that an acquisition
of dignity of character was derived from it. He expressed a
particular enthusiasm with respect to visiting the Wall of China.
I catched it for the moment, and said I really believed I should go
and see the Wall of China had I not children, of whom it was my
duty to take care. 'Sir (said he), by doing so, you would do what
would be of importance in raising your children to eminence.
There would be a lustre reflected upon them from your spirit and
curiosity. They would be at all times regarded as the children of a
man who had gone to view the Wall of China. I am serious, Sir.'

Though Boswell never saw the Shanxi mountains or the Yellow River, his
friend's confidence was not misplaced. Stretching from Chinese Turkestan
to the Pacific in a line of well over 3000 kilometres, the Wall covers a
distance getting on for one-tenth of the Earth's circumference. To visualise
its equivalent in Europe, one must think of a continuous structure reaching
form London to Leningrad. The Roman Empire had its frontier fortifications covering stretches between rivers and other natural borders, but they
never attained anything like this length; the longest, connecting the Rhine
and the Danube, did not exceed 560 kilometres.

There is no lack of travellers' descriptions of the Great Wall, and Fig. 400
gives a good impression of it winding over the mountains of Hubei and
Shanxi. Those who have walked long distances along it estimate that it has
some 20 000 wall-towers and that 10 000 watch-towers are still standing,
but at the time of maximum strength the Wall would have had at least 5000
more of each of the two types.

The component parts of the Wall are sketched on the map (Fig. 390). Its
present main line is, broadly speaking, that of the Ming period, but different
parts of it are built along alignments of earlier walls dating from very different periods. The eastern end begins at Shanhaiguan (101), where one of the
portals on the road to Manchuria and Korea bears the inscription *Tianxia
di Yi Guan* (The World's First Gate), then, as it moves westwards, the Wall
ascends steeply on to the mountains overlooking the Hubei plain. Next it
snakes along ridge after ridge in a section broken by only one important
gate, that of Gubeikou (108), through which passed a road so important for
the imperial summer residence during the Qing. Almost due north of Beijing
(7), the Wall divides in two: the Outer, or Northern, Wall running along the
border of Shanxi province; the Inner Wall descending to a point 200 kilometres further south and then rejoining the main defence line almost 50
kilometres east of the Yellow River. It is in this Inner Wall, at the Nankou
Pass (109), that there is the famous Zhuyongguan Gate, through which one

Fig. 400. The Great Wall near the Nankou pass, north of Beijing. Note one of the slotted access staircases in the right-hand lower corner. (Photo. from Jin Shousen, 1962.)

reaches the important cities Zhangjiakou (110) and Datong (86), both giving access to Inner Mongolia.

The Outer, or Northern, Wall is, in a sense, comparatively late, for it is undoubtedly based on an alignment dating from the fifth and sixth centuries AD. After provisional consideration during the Northern Wei, the first construction was undertaken by the Eastern Wei in AD 543, then carried on more energetically by the Northern Qi between AD 552 and 563, especially in a great effort covering around 1500 kilometres in AD 556, which cost heavily in men and materials and almost bankrupted the State. The Inner Wall is of uncertain date, but a section of some 370 kilometres along the edge of the Shanxi uplands overlooking the Hubei plain is certainly somewhere near the site of a wall built in the fifth century BC by the feudal State of Zhongshan. The rest is probably of much later date, perhaps dating from about AD 390, or alternatively from the Wu Dai period (AD 907 to 960).

Once across the Yellow River, the Wall runs south-west across the Ordos Desert, then rises north-west to meet the river near Ningxia. The first part is very ancient, having been constructed by the State of Wei in 353 BC. In the neighbourhood of Lanzhou (27), the layout again becomes complex. There are remains of a wall for a long way following the right bank of the river, then, from Lanzhou north-westwards, a wall in better preservation protects closely the Old Silk Road, but, in addition, an outer wall strikes off across desert country to rejoin the original wall in the neighbourhood of Liangzhou (30). The dates of these modifications are uncertain. North-west of Liangzhou, the line continues as earthworks with periodical stone towers until it reaches the fortress of Jiayu Guan (36) at the Western Gate, and it ends a few kilometres further in towards Qilian Mountain, to form what in Ming times and for many centuries earlier was known as the 'Last Gate of the World'. However, this was not the case in the second century BC (during the Han), for at that time a further earthwork wall standing at least 3.6 metres high and dotted with many 9 metre towers continued past Anxi (38) to surround Dunhuang in a protective embrace.

A very mysterious extension of the Great Wall exists in the form of the 'Qinghai Loop' in the neighbourhood of Lanzhou (27). Originating from the western junction of the Lanzhou Loop with the Outer Wall, it passes south-west in an arc enclosing Xining (115), crosses the Yellow River and goes back to Lanzhou, but has a short branch off to the south. This seems to be an alignment dating back to the fourth century AD when the Tuguhun tribes were a great menace in the Tibetan hills to the west.

Thus, in sum, the Wall has had many periods of importance and periods of decay. After the third century AD, there was little maintenance, but during the sixth and seventh centuries AD there was major reconstruction. However, from then on, for seven centuries during the Tang and Song, the wall

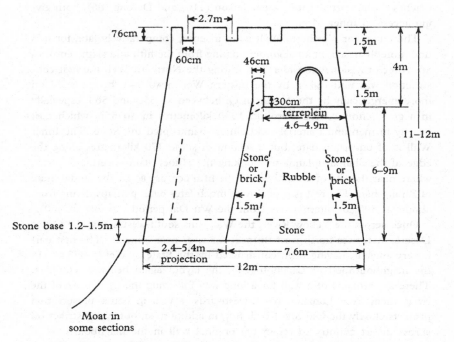

Fig. 401. Approximate dimensions of wall and towers in the eastern stretches of the Great Wall.

received neither upkeep nor fresh building. Then, during the Yuan, the Wall lost all significance, and this is why its present state is essentially Ming, for the following dynasty, the Qing, also found it irrelevant.

The approximate dimensions of the wall and towers in the eastern stretches along the Hubei and Shanxi borders are shown in Fig. 401. The granite blocks of the stone foundations are often as large as 4.3 metres × 0.9 or 1.2 metres, and those of the stone facings for the rubble core some 1.5 metres × 0.6 metre × 0.45 metre; if the facing (always about 1.5 metres thick) is of brickwork, it contains seven or eight thicknesses. There are five to seven or eight towers per kilometre, at distances ranging from 90 to 120 metres. To compare the constructional methods of past ages with those which would be employed now would lead us far into conjecture. One can only say that an immense amount of man-handling of the blocks on slides must have been used, with suitable tackle for laying it in place.

Reinforcements of wood, and even of iron, appear to have been used occasionally in certain sections of the wall. The wood was very hard, and in the thirteenth century AD, when parts broke after heavy rain, it was used for spear-shafts. It may have been shuttering, but probably the baulks were intended as reinforcement, for Han forts in the Tarim basin were generally formed of long cylindrical bundles of brushwood and wild poplar trunks alternating with layers of tamped clay. On the other hand, it could have been piling; in the seventeenth century AD, Fang Yi-Zhi wrote that if piles of '1000-year wood' were rammed down, 'much energy is saved; pine and cypress wood can last for centuries without decay'.

There is evidence that the first Great Wall, that of Qin Shi Huang Di, took quite a different course from the present main line. How far north-west of Lanzhou it passed is unknown, but it certainly passed Ningxia (114) and then kept all the way north of the great bend of the Yellow River, covering Wuyuan (5), where there was a gate at Gaoque (117), a site long lost. The surmise is that it then ran eastwards through the southern steppes of Inner Mongolia, along a line some distance north of the present Wall, reaching the sea not far from Shan Hai Guan (101), the Eastern Gate. There is also evidence that the Qin fortifications were extended, probably as an earth-work, so as to reach the sea again near the mouth of the Yalu River, the present frontier between China and Korea. A little more information comes from the biography of the Qin general Meng Tian:

> After Qin had unified the world (in 221 BC), Meng Tian was sent to command a host of 300 000 men to drive out the Rong and the Di (barbarians) along the northern (marches). He took from them the territory to the south of the (Yellow) River, and built a Great Wall, constructing its defiles and passes in accordance with the configurations of the terrain. It started at Lintao (119) and extended to Liaotong, encompassing a distance of more than 10 000 *li* (almost 5000 kilometres). After crossing the (Yellow) River, it wound northwards, touching Mount Yang.

But of the immense organisation which the task must have involved, of the supply trains, of the surveying and planning, no word has come down to us.

A work planned on so vast a scale seems to have raised superstitious fears of undue interference with the given pattern of Nature. Some suggested that in places it had 'cut through the veins of the Earth', and a story even grew up that Meng Tian was ordered to commit suicide. There are good reasons for thinking that the story is a literary invention, but it has an intrinsic interest, as it may well have some connection with ancient engineering controversies about the proper ways of dealing with watercourses, a subject which will be discussed later (on pp. 172 ff.).

The building of the Great Wall can only be viewed in its correct historical perspective by realising that it was not so much the construction entirely from scratch of a continuous work, but rather the linking of a number of

Walls which had been built previously by the various Warring States, the purpose of which was to break the shock tactics of nomadic horse-archers, or of cavalry belonging to feudal States which had adopted such tactics.

Pre-Qin walls are sketched in Fig. 402. Their construction began in the late fourth century BC with a number of defence works against nomads from the steppes. Around 300 BC, the Qin State built one wall from the Tao River in Gansu north-eastwards to somewhere near the place in northern Shanxi where the present Great Wall turns northward again. Thus it joined an earlier Wei wall of 353 BC which ran further north-eastwards along the edge of the Ordos Desert towards the descending loop of the Yellow River, a line more or less followed nine centuries later by the Sui Great Wall. Also around 300 BC, the State of Zhao built another wall running from Gaoque (117) eastwards north of the Yellow River to somewhere between modern Kalgan and Beijing. In 290 BC, the State of Yan built a third fortification from near the eastern end of the Zhao wall to the lower valley of the Liao River in Manchuria. Both these were very near, if not exactly on, the line taken by the Great Wall more than half a century later. But the walls were not only to keep out the barbarians or to prevent marginal Chinese from joining them, for some were built along the boundaries of individual feudal States. Yet one cannot suppose that these social entities permitted construction of much more than a continuous dyke and ditch to aid defence; probably these early walls were something like the series of Romano-British 'Devil's Dykes' between the ancient fen and forest in East Anglia. It is very unlikely that they were faced with brick or stone as were large sections of the later Great Wall.

The best estimate of the Great Wall's length puts it at about 6324 kilometres, if all its branch walls are counted, and some 3460 kilometres if the main line alone is taken. These figures, which are broken down in Table 54, are really remarkable when the relatively primitive state of transportation in the Qin and Han is taken into consideration.

The effectiveness of the Great Wall in keeping out the troops of nomadic horsemen was probably considerable. Any breaking down of the wall, or the building of ramps up to it, would allow time for the arrival of Chinese reinforcements. Indeed, the impenetrability of the Great Wall certainly seems to have been a factor in starting a series of shocks in tribal relations which transmitted themselves like a chain reaction to give rise to disturbances and invasions of the nomad settlement frontier in Western Europe. Chinese engineering skill, and the genius of the people at that time for the organisation of mass labour projects, might therefore be said to have outplayed the protective abilities of the Roman Empire. For while the Romans could build Hadrian's wall, which stretched for 118 kilometres and was 2.5 metres thick, across the narrow neck in northern England in the second century AD,

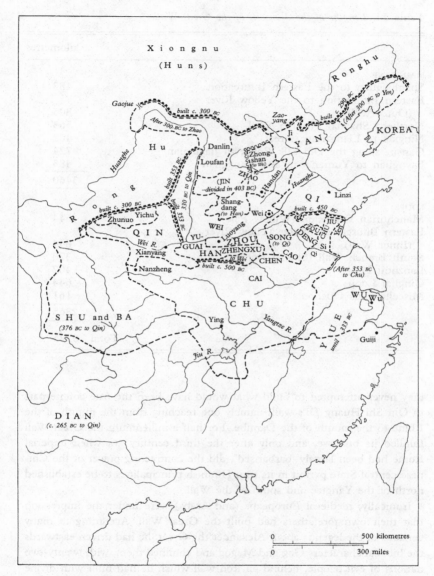

Fig. 402. Approximate boundaries of the Feudal States and the layout of the Great Wall about the beginning of the third century BC. Based on Herrmann (1935), by permission.

Table 54. *Lengths of the divisions of the Great Wall.*

	kilometres
MAIN LINE	
Shanhaiguan to the Eastern Bifurcation	483
Eastern Bifurcation to the Yellow River (Outer Wall)	805
Northern Shaanxi Frontier Wall	563
Ningxia to Liangzhou (direct)	402
Gansu, along the Old Silk Road, to Jiayuguan	724
Jiayuguan to Yumenguan and Yangguan	483
	3460
LOOPS	
Manchurian Extension (Willow Pale)	644
Eastern Bifurcation to the Yellow River (Inner Wall)	644
South Branch Wall (Shanxi Border)	370
Lanzhou Loop	402
Qinghai Loop	644
Miscellaneous Offshoots	161
	2865
Total	6325

they never attempted to build what would have been the true counterpart to Qin Shi Huang Di's wall, namely one reaching from the mouth of the Rhine to the mouth of the Danube. For half a millennium the Great Wall fulfilled its purpose, and only after the third century AD, when imperial Rome had been largely 'barbarised', did the contracting power of the Chinese central State permit in its turn Hunnish principalities to be established north of the Yangtze and south of the Wall.

Ironically, medieval Europeans (and Arabs) were under the impression that their own forefathers had built the Great Wall. According to many versions of the legends about Alexander the Great, he had driven eastwards the biblical characters Gog and Magog and confined them, with twenty-two nations of evil people, behind an iron wall which he had built with divine assistance. In the last days, they would break through his gate and overrun the world. Yet by the time that His Excellency Ysbrants Ides, Ambassador from His Tsarish Majesty to the Emperor of China, rode with his cavalcade through the Nankou Gate in October 1693, Europeans well knew who deserved the credit.

3

Building technology

Though architecture is a subject which lies so near to the fine arts that it hardly comes within the scope of this book, it has a technological basis which cannot be omitted. Ceramics is just such another case. In consequence, we can only glance here at the abundant literature concerned with the aesthetic aspects of Chinese civilisation.

To the first European visitors in the sixteenth and seventeenth centuries AD, Chinese buildings must have seemed very strange. They attracted more detailed interest in the Chinoiserie period of the eighteenth century, yet even so, by 1757, when *Designs of Chinese Buildings* by Sir William Chambers was published, the principles of Chinese building construction were only half understood. Things were not much better during the nineteenth century, when only two books appeared in English, but more recent studies have partially made up for previous neglect. All the same, after some time a proportion of readers will begin to feel that there is a surfeit of beautiful photographs and too much archaeology and comparative religion, and would prefer more precise information about the functional basis of the construction of buildings in China. There is also a feeling that less study should be given to the highest flights of Chinese architecture, and more to the regular, commonplace, but regionally diverse and often very attractive, dwellings of townspeople and the farming community. Again, looking at so many examples of the exquisite curving roofs, there is the desire to have clearer ideas about the history of the remarkable style which produced them. To these needs there has been only a partial response in the West.

THE SPIRIT OF CHINESE ARCHITECTURE

In no other field of expression have the Chinese so faithfully made manifest their great principles, first that the human race cannot be thought of apart from Nature, and secondly that mankind cannot be separated from its social aspects. Not only in the great constructions of temples and palaces, but also in domestic buildings scattered as farmsteads or collected in villages and towns, was there embodied throughout the ages a feeling for a cosmic pattern and the symbolism of the directions, the seasons, the winds and the constellations. Preferring to construct all buildings for the living in the relatively impermanent media of wood and tile, bamboo and plaster, the Chinese made the use of horizontal spaces the keynote of their architecture and planning. Though some buildings might be of one or two storeys above a ground floor, their height was strictly subordinated to the large-scale horizontal perspectives of which they might form part. Significantly, the pagoda, the Chinese form of the vertical spire, retained to the end the apartness of its foreign origin, and continues to grace isolated hills outside cities, yet, though removed from the architectural whole, it is still related, as nothing could fail to be in China, to the landscape. For the quality of all the planning in Chinese architecture was always with, not against, Nature. It did not spring suddenly out of the ground as if aiming to pierce the sky like a European 'Gothic' building, nor did it force trees and plants into Renaissance straight lines, lozenges and triangles. It took every advantage of the natural beauty of the site, of woods and hills; and this is true not only of such marvels as the Summer Palace near Beijing, but also of the ordinary Sichuanese farmhouse surrounding its threshing-floor and backed by a clump of bamboos, at the head of its valley of terraced rice-fields (Fig. 403).

The fundamental concept of Chinese architecture lies in the arrangement of one or more courtyards to compose, sometimes in a very complex way, a general walled 'compound' (Fig. 404). The main lengthwise axis is always (or ideally) north–south, and the chief buildings, or halls, are always placed at right-angles to it. They thus come one behind another, with the main entrance of each always in the centre of the long south-facing side. These rectangular buildings are sometimes connected by means of open galleries, variously planned, with rows of smaller buildings flanking the courtyards on both sides. With this system, enlargement is never carried out by adding to the height, but by continual duplication of existing units, and growth in breadth or preferably depth. The entrance to a large composition is likely to be through a gate, which may resemble a city gate in having a barrel-vaulted way below and a pavilion above, or simply cap parallel tunnels with a heavy roof. The halls may occasionally be of two storeys, and small open pavilions or pavilions of more than two storeys may be used, often away from the main axis, for the diversification of the general plan (Fig. 405). In domestic buildings, the courtyard system has the great merit that yards and gardens are made part of the building, and are not something additional or separate. The spaces between the low houses supply abundant air, and all the windows look out internally on to plants and trees in the garden courtyards. Thus people are not isolated from Nature.

Fig. 403. A typical farmhouse, near Shaoshan in Hunan. (Photo. Joseph Needham, 1964.)

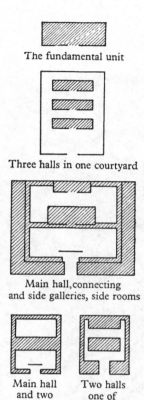

The fundamental unit

Three halls in one courtyard

Main hall, connecting
and side galleries, side rooms

Main hall
and two
courtyards
with side
rooms

Two halls
one of
which
stands in
isolation

Fig. 404. Typical ground plans of
Chinese buildings.

The long rear wall is almost always unbroken by doors or windows, and
forms, as it were, the ultimate statement of the plan, though not its climax,
since the largest hall will be placed somewhere north of the central point,
and there will be a diminuendo of constructions behind (i.e. to the north of)
it. For a concrete illustration in the Chinese style of draughtsmanship, we
may glance at a plan of a Confucian temple (Fig. 406) taken from the *Sheng
Xian Dao Tong Tu Zan* (Comments on Pictures of the Saints and Sages,
Transmitters of the Dao) of AD 1629. Here may be seen the triumphal
gateways, the succession of halls and avenues on the central axis, the arrangement of two subsidiary parallel axes, the enclosing walls and various
pavilions for special uses. Even domestic architecture, moreover, has an
informally liturgical character, which reflects the ancient prescriptions of the

Fig. 405. Central octagonal pavilion, sited on the main axis of the 'Blue Goat Temple' in Chengdu, Sichuan. (Photo. Joseph Needham, 1943.)

books of social ceremonial. This is well illustrated in Fig. 407, which gives a bird's-eye view of a traditional home in Beijing.

Another characteristic of the Chinese house, seen as soon as it rises above the level of importance of the simple thatched cottage, is that it is based upon a platform. From the earliest times, this was probably a utilitarian expedient to raise the living quarters and the passages between them above the mire of the farmyard and visiting caravans. With the passage of time, this developed into one of the most majestic elements of the full Chinese style, joining with others, such as the great emphasis placed on roofs and the advantage invariably taken of sloping ground. A result of the last is that, in some large buildings, the main entrance is at the lowest point, so that the visitor wanders through the series of courtyards and halls ever ascending. Indeed, in very important buildings, the terrace platforms may exceed 5 metres in height, and may be built of white marble, access to them being by two central stairways flanking a central inclined 'spirit path' carved in high relief, besides which there will also be staircases at the sides of the courtyard.

Sirén has commented 'The wooden pillars rise above the supporting

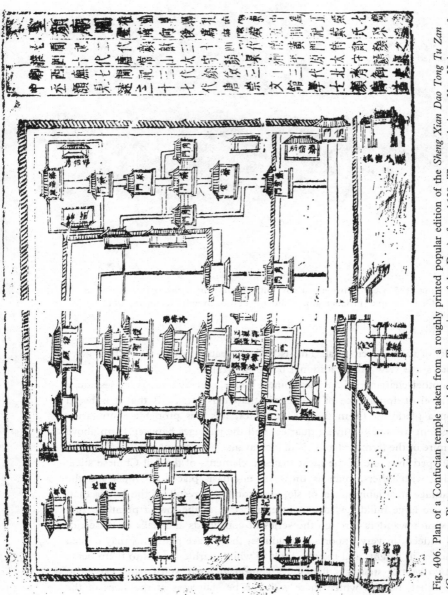

Fig. 406. Plan of a Confucian temple taken from a roughly printed popular edition of the *Sheng Xian Dao Tong Tu Zan* (Comments on Pictures of the Saints and Sages, Transmitters of the Dao) of AD 1629. The temple is to Yan Hui, one of the 'four associates' of Confucius, and his favourite pupil; it is situated at Qufu, east of the temple of Confucius himself. Symmetry dominates the placing of the subsidiary fasting pavilions and the sacrificial shrines of the sage's distinguished descendants. The entrance is first through the triumphal gateway and then through the ordinary gateway at the bottom. The text on the right mentions some of the men whose lesser votive altars are within the compound, for example on the east side.

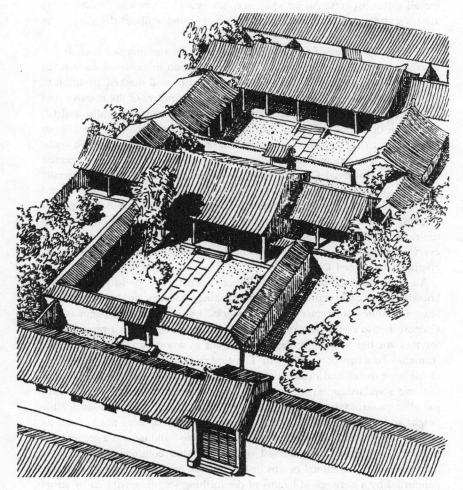

Fig. 407. Sketch of a traditional Chinese home in Beijing. (After S. E. Rasmussen.)

terraces, which often reach considerable heights, like tall trees on mounds and hillocks. The lines of the far-projecting curving roofs suggest the long wavering branches of cryptomerias, and if there are any walls, they almost disappear in the play of light and shade produced by the broad eaves, the open galleries, the lattice-work of the windows, and the balustrades.' This point is important, for the walls of Chinese buildings are indeed of secondary significance; it is the terraces and the overhanging roofs which are decisive for all outer aspects. As we shall see, the walls of Chinese buildings are always curtain walls; they take no share as bearing walls in the support of the structure.

The main features of Chinese buildings may be summarised then as follows: (a) emphasis on the roof, and its construction in sweeping curves; (b) formal grouping of buildings round central courts, and marked attention to axis; (c) frankness of construction, the supporting pillars of the massive roof timbering being clearly visible, even when partly engaged in walls; and (d) lavish use of colour, not only in roof tiles, but on painted columns, lintels and beams, richly bracketed cornices, and broad expanses of plastered walls. But as the Western architect Andrew Boyd has said, the quality of greatest interest in traditional Chinese building is perhaps that it was functionally and structurally direct and honest. The structural elements are distinct and explicit, all decoration being based on them. Clarity and rationality appears in plan, section and elevation, and in the high degree of harmony between the three. 'Chinese buildings, for all the sophisticated aesthetic that controls every part, have a look of being built by a master-craftsman or architect-engineer, as indeed they were.'

All was under the aegis of the Chinese element Wood. In spite of early knowledge of constructing arches and vaulting, masonry and brickwork were always confined to terraces, defence works, walls, tombs and pagodas. No Chinese house could be a proper dwelling for the living, or a proper place for the worship of gods, unless it was built in wood and roofed with tile. Immense consequences followed. The timber frame and curtain walls provided large spans, and maximum unobstructed space, as well as flexibility of use and standardisation in planning. The timber structure blossomed into the elaborate roof, which became the building's main feature; considerable height and monumentality, when occasion demanded, were not beyond its powers. As we shall see, the most typical Chinese building was a rectangular hall on a terraced base, marked out by wooden columns in a complex construction of horizontal beams or lintels. The heavy overhanging roof was supported by a network of beams of diminishing length, separated by struts that were superimposed one over the other across the rectangle. Since the horizontal members of the roof – the purlins – were supported by this network, any desired curve could be given to the roof. All round the building the beams were cantilevered outwards to form the generous overhang of the eaves, and as time went on there arose an extremely elaborate development of cantilever brackets piled up in tiers, to reach a maximum overhang in Tang and Song times, and embodying the most ingenious carpentry. To this must be added flexibility for expansion in all directions and marvellous polychrome ornament provided by the decorators.

As far as 'standardisation of planning' is concerned, it is worth remarking that modern architecture has been influenced more by Chinese (and Japanese) ideas than is usually supposed. One basic Chinese characteristic has always been the addition at will of repeating units keyed to the size and scale of human beings – pillar intervals, or bays in buildings, and spaces in open-air courts. Such modules occur in the theory and practice of modern architects, such as le Corbusier; indeed, Frank Lloyd Wright worked in Japan, while H. K. Murphy laboured in China.

In China, traditional architects and builders were extremely conscious of standard dimensions and correct proportionality. Indeed, in the *Ying Zao Fa Shi* (Treatise on Architectural Methods) of AD 1097, a particular proportion was used. This was the end elevation of the horizontal corbel bracket arm (see p. 68), which gave rise to one *cai* [*tshai*], and an area of 2 × 1 *dou-kou* [*tou-khou*], the *dou-kou* being a relative dimension. By the tenth century AD, this ranged in twelve different sizes from 15 centimetres down to 2.5 centimetres, though in the Song the unit was slightly larger.

Although the term *cai* [*tshai*] had this special technical significance, its use was certainly related to the fact that the timber beams and baulks came in a variety of predetermined sizes, also called *cai*. 'The height and thickness of the corbel brackets must always correspond with (those of) the standard timbers', says the *Treatise on Architectural Methods*. It goes on:

> As for the roof-structures of houses and halls, everything depends on the standard size of the materials chosen. There are eight standard sizes (of the height and thickness of timber baulks), and these are used in accordance with the size of the building. [Measurements follow, ranging from some 23 centimetres × 15 centimetres to about 11 centimetres × 8 centimetres. Notes specify the type of buildings, partly in terms of numbers of rooms, for which each grade of cross-section is suitable.]
>
> When the measure called *qi* [*chhi*], six standard parts in height and ten in thickness, is placed on top of the *dan-cai* [*tan-tshai*], the whole is called the *zu-cai* [*tsu-tshai*].
>
> The height of each standard beam is divided into fifteen standard parts, and the thickness (always) corresponds to ten of these.
>
> Thus the height and depth of the roof, the length, curvature and trueness of the members, and the ratios of column and post heights (lit. raising and cutting) (in the structural cross-section adopted for any particular ground plan), together with the right use of square and compass, plumbline and ink-box – all proportion and rule depends on the system of standard timber dimensions and the standard divisions of these.

This system was undoubtedly inherited from Tang practice, and was probably not new even then. Thus the entire Chinese building was designed in terms of standard modules and modules of variable absolute size, none ever out of scale with human proportions. Right proportion was thus safeguarded, and relational harmony preserved, whatever the magnitude of the structure.

To unravel the way the various building patterns of China were determined by the needs of different parts of the social order at different times is a special task in itself, and an important one, though here we can make only a few remarks. For example, the large joint-family system, in which the sons did not quit the ancestral compound on marriage, must have given rise to differentiation between a multiplicity of halls and courtyards within a single enclosure. So also the tendency of great families in the Han to establish what were almost factories in their dwellings. After the Tang, the wealthy family was not so often a centre of production of commodities, but then the effects of production among the families of artisans on urban buildings became evident, especially in Chang'an and Hangzhou, as well as in the great cities of subsequent times.

Rural building must always have centred round the farm as an agricultural production unit, and domestic architecture must have waxed and waned in accordance with the prosperity of countrysides and the particular social strata (poor peasants, rich peasants, scholar-gentry, etc.) in question. As for the formative influence of the bureaucratic-feudal State on Chinese public architecture, it was evidently capable of large-scale planning from the start, and its splendid works were essentially secular in spirit. Confucian attitudes emphasised the ethical, hierarchical, axial and symmetrical qualities, with a certain inwardness too. This last was echoed by Daoist influence, though it generally promoted softer, less severe, qualities, beautifully sited romantic ensembles and the development of the garden and the artificial landscape. Buddhism went along with Daoism, adding, however, the pagoda derived from the Indian *stūpa* or memorial shrine (see p. 105), as well as the triple- or five-span gateway. The wall of the enclosure turned into long cloisters facing inwards, and, while the gates held their place, the pagoda, at first central, was duplicated symmetrically, or pushed to one side and finally exiled to the outer grounds. North of its primitive mid-line position came the worship hall, north again the lecture hall, and, still further north, the dormitories and living quarters.

Essentially, however, there was never any dividing line between the secular and the sacred in East Asian architecture. Indeed, palaces were sometimes converted to temples, and temples to schools or hospitals. Perhaps all this was an echo of the fundamentally organic and integrated quality of Chinese thought and feeling.

Whatever the forces which moulded the Chinese building trade, its achievements were truly extraordinary. It mirrored in hard structural materials the outstanding genius of this people for combining the rational and the romantic. Harmonising intellect and emotion, it wedded the science of the erection and disposition of buildings to the art of landscape design in such a way that Nature remained dominant, free from subjection to an imposed architectural pattern, and rather uniting with the human works in a larger synthesis.

The planning of towns and cities

If the individual family dwelling, the temple or the palace was so elaborately and attentively planned, set out indeed as an organic pattern, it would naturally be expected that urban planning would also show a considerable degree of organisation. The question is not, however, quite so simple, for in China there was a rather marked difference between the spontaneously growing village or rural settlement and the town planned from the start. Villages tended to grow up, as elsewhere in the world, along roads, paths and other lines of communication; they arose at crossroads, or where three ways met. At all times, they possessed a great deal of unofficial self-government; the sense of community was very real, especially when the village was far away from lines of communication in order to be convenient for working an agricultural area surrounded by upland or forest.

The Chinese town, on the other hand, was not a spontaneous accumulation of population, nor of capital or facilities of production, nor was it essentially a market-centre. Above all, it was a political nucleus, a local centre in the administrative network, and the seat of the bureaucrat who had replaced the feudal lord. Originally, before the first century BC, the feudal chieftains appropriated the centres of assembly where the people exchanged commodities and came together for the seasonal festivals. There was therefore no distinction throughout Chinese history between the feudal castle and the town; the town *was* the castle, and was built so that it could serve as the protection and refuge, as well as the administrative centre, of the surrounding countryside. Towns and cities in China were not the creation of burghers, and never achieved any independence with respect to the State. They existed for the sake of the country and not vice versa. They were planned as rational fortified patterns imposed from above on carefully chosen places. However, they did not necessarily grow; indeed, they shrank as often as they expanded, their walls remaining to be used again perhaps if a town was required during a later dynasty. Their population was merely a sum-total of individuals, each of whom was closely linked with the village from which the family had originated, and where its ancestral clan temple still stood. While the European city developed from within outwards, centred on its market-place, cathedral, forum etc., the external fortifications were the essence of

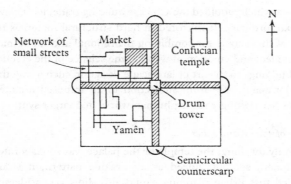

Fig. 408. Typical city plan. The yamên is the governor's residence.

the Chinese city, and the key points were the drum-tower and the offices of the governors, civil and military.

It is probable that all Chinese cities were laid out, since the Zhou period, in a rectangular manner, thus closely resembling the Roman fortress. There was the great east–west street and the great north–south one cutting it at right-angles. Indeed, some have thought that certain ancient forms of Chinese characters betray an older circular type of wall, but the arguments assume that in the difficult writing media of bone and bronze the Shang and Zhou scribes were really capable of distinguishing clearly between the round and the square, which may be doubted.

In any case, the most typical form of a Chinese city is shown in Fig. 408. The palace of the feudal lord, and afterwards the *yamên*, or headquarters, and residence of the civil governor, was usually in the 'front' or southern part, with the market-places situated more to the north. From the main crosswise plane, there always followed a rectangular pattern of streets, dividing the city into blocks. Often these were separated by walls and gates, constituting areas under the authority of a subsidiary of the city magistrate. Later developments in Sui and Tang times led to a concentric series of walled enclosures, the palace innermost, then the administrative offices, and finally the people's city with its external walls.

Even up to the present day, the principal walls of cities in China often form a square or rectangle, though there are many exceptions. The very long walls of Nanjing follow the local topography, and some large cities, such as Fuzhou, have very irregular outlines. Occasionally there occurred a circle or ellipse, as in Shanghai in Song times. Nearly all cities still have wide empty spaces left within the walls, available for kitchen-gardens and even farms. Sometimes ribbon-development along an important road outside the city gates became incorporated later within the walls, forming an elongated extension, as is seen in Lanzhou. Another city of Gansu, Tianshui, came thus to consist of five walled towns joined together in a row.

In Chinese cities, a rather large population was sometimes packed into a confined area. In general, the builders did not resort to multiple storeys; instead, party walls were constantly used to separate dwellings of different families, and even the wealthy had rather restricted space, but every court-yard, however small, became something of a garden by the use of plants and small trees in pots. This meant that population density could reach high levels and yet a sense of seclusion was preserved. In Beijing, population figures for residential areas exceeded 21 000 per square kilometre, and for working areas it was nearly 33 000 per square kilometre, but the city main-tained a garden character owing to the abundance of trees, which, paradoxi-cally, were more numerous within the walls than outside.

To sum up, then, we find a series of distinct spaces, each opening into the next but screened from it by walls, gateways, buildings overhead, and at chosen points such 'incidents' as, perhaps, a bow-shaped stream with its marble balustrades. Between the constituent parts there is remarkable bal-ance and interdependence.

The contrast between a typical Chinese city and the Renaissance palace is striking, for in the latter the open vista, as at Versailles, is concentrated upon a single central building; the palace is something detached from the town. The Chinese conception was much grander and more complex, for in one composition there were hundreds of buildings, and the palace itself was only part of the larger organism of the whole city with its walls and avenues. Although so strongly axial, there was no single dominating centre, but rather a series of architectural experiences. The Chinese conception invites a dif-fusion of interest. The whole length of an axis is not divulged at once, but rather a succession of vistas, none of which is overpowering in scale. Thus the Chinese form of the great architectural ensemble, which had attained its highest level already in the early fifteenth century AD with the Altar and Temple of Heaven at Beijing (Fig. 409), combined a meditative humility attuned to Nature with a poetic grandeur to form organic patterns unsur-passed by any other culture.

Building science in Chinese literature

Presumably owing to the fact that architectural employment was not consid-ered a very suitable occupation for a Confucian scholar, Chinese literature is relatively poor in writings on the subject. However, the earliest dictionary, the *Er Ya* (The Literary Expositor), dating from Zhou and early Han times, has a special chapter devoted to special matters connected with building. In this we find a good many technical terms which retained their meanings

Fig. 409. Aerial view of the sacred domain (*temenos*) of the Altar and Temple of Heaven in Beijing, seen from the south. In the foreground are the Orbed Concentric Platforms of the Altar, then the Hall of the Infinite Canopy of Heaven (the smaller round building) and, at the northern end of the causeway, surmounting platforms round and square, the Hall of Prayer for the Year. To the right at the top is the complex of buildings for the fasting and preparation of the imperial celebrant. (Photo from Anon. (1849).)

afterwards with little or no change. Later encyclopaedias often have similar sections. In Qing times, a number of scholars carried out useful studies to elucidate the meaning of ancient architectural words and expressions.

The chief literary tradition which involved actual architectural plans was that of the *San Li Tu* (Illustrations of the Three Rituals). Two books of this name were written in the Han period; subsequently they were combined, and in AD 600 an important series of illustrations was added. In or around AD 770, a further revision was made. All this work was afterwards lost, but not before some of it had been used in a definitive text of AD 956 by Nie Chongyi, though a little more than a century later the astronomer, engineer and high official, Shen Gua, expressed doubts as to whether the pictures then existing could be considered authoritative. The text of Nie has come down to us, re-edited for the last time in 1676 by the Manchu prince Nalan Chengde.

The reason why this material was preserved even as well as it was lay in the desire of scholars to interpret sections of the ancient liturgical and ceremonial texts. The plans of architectural interest include the Ming Tang (cosmic palace-temple), Gong Qin Zhi (public halls and domestic apartments) and the Wang Cheng (princely city) (Fig. 389). At various times, efforts were continued to throw further light on the buildings which must have been visualised by the writers of the Rituals; the planning of imperial ancestral temples was a related study. All this, however, was on the purely scholarly level, concerned with the general layout rather than with the techniques of construction, and somewhat remote from the worlds of practical architects and building workers.

Nevertheless, practical architects also had a literary tradition. From early times there must have been manuals of procedure for the various trades carried on in, or under, the auspices of the imperial workshops. The emperor Yuan of the Liang dynasty (AD 502 to 557) tells us that in the third century AD there were elaborate rules for the erection of all kinds of buildings. But not even the titles of such manuals have been preserved in the official lists, and even the most important of them would not have survived if it had not been mentioned by certain Song writers. This was the *Mu Jing* (Timberwork Manual), the authorship of which was attributed to a famous builder, Yu Hao, who flourished at the beginning of the Song during the decades AD 965 to 995. He constructed the Kaibao pagoda at Khaifeng, which was universally regarded as a marvel of art but was struck by lightning and destroyed about AD 1040. The fact that his book was not recorded in the official bibliography is significant because it shows that building technique was regarded as too 'mechanical' for inclusion among scholarly works. There was probably also a social barrier, for Yu Hao was a Master-Carpenter, while the man who built upon his work to produce the greatest architectural book in Chinese history was a 'white-collar' Assistant in the Directorate of Buildings and Construction.

Shen Gua wrote about Yu Hao in a passage which is worth quoting in full. He said:

> Methods of building construction are described in the *Timberwork Manual*, which some say was written by Yu Hao.
> (According to that book), buildings have three basic units of proportion; what is above the cross-beams follows the Upperwork Unit, what is above the ground floor follows the Middlework Unit, and everything below that (platforms, foundations, paving, etc.) follows the Lowerwork Unit.

The length of the cross-beams will naturally govern the lengths of the uppermost cross-beams as well as the rafters, etc. Thus for a (main) cross-beam of 2.4 metres length, an uppermost cross-beam of 1 metre length will be needed. (The proportions are maintained) in larger and smaller halls. This (2.4) is the Upperwork Unit.

Similarly, the dimensions of the foundations must match the dimensions of the columns to be used, as also those of the (side-) rafters, etc. For example, a column 3.3 metres high will need a platform 1.2 metres high. So also for all the other components, corbelled brackets, projecting rafters, other rafters, all have their fixed proportions. All these follow the Middlework Unit (2.75).

Now below of ramps (and steps) there are three kinds, steep, easy-going and intermediate. In palaces these gradients are based upon a unit derived from the imperial litters. Steep ramps are ramps for ascending which the leading and trailing bearers have to extend their arms fully down and up respectively (ratio 3.35). Easy-going ramps are those for which the leaders use elbow length and the trailers shoulder height (ratio 1.38); intermediate ones are negotiated by the leaders with downstretched arms and trailers at shoulder height (ratio 2.18). These are the Lowerwork Units.

The book (of Yu Hao) had three chapters. But builders in recent years have become much more precise and skillful than formerly. Thus for some time past the old *Timberwork Manual* has fallen out of use. But (unfortunately) there is hardly anybody capable of writing a new one. To do that would be a masterpiece in itself!

This passage would have been written about AD 1080. Within twenty years the man capable of doing the job which Shen Gua saw was necessary had arisen and completed it. This was Li Jie, the author of the *Treatise on Architectural Methods.*

The date of Li Jie's birth is uncertain, but he was already a subordinate official in the Bureau of Imperial Sacrifices when Shen Gua was about to produce his *Meng Qi Bi Tan* (Dream Pool Essays) in AD 1086. Moving to the Directorate of Buildings and Construction in 1092, he must have shown immediate and outstanding promise as an architect, for his revision of the old treatises was commissioned in 1097, completed by 1100, and printed three years later. He was a distinguished practising builder as well as a writer, for he erected administrative offices, palace apartments, gates and gate-towers, and the ancestral temple of the Song dynasty, as well as Buddhist temples. Li Jie says in his preface that he studied long and minutely the practices and orally transmitted rules of the master-carpenters and other responsible artisans. Yet it is of much interest that he never quite succeeded in fusing the scholarly and the technical traditions.

However, Li does give the Rules and Regulations, and these form the main body of the book. They deal systematically with one department after another, comprising:

Moats and fortifications	Turning and drilling	Painting and decorating
Stonework	Sawing	Brickwork
Greater woodwork	Bamboo work	Glazed tile making
Lesser woodwork	Tiling	
Wood-carving	Wall building	

The last chapters deal with job accounting, materials (including some interesting paint compositions) and the classification of crafts. From the first, the book had contained excellent illustrations (see Figs. 410, 413 and 414). The attention given to the basic construction and the shaping of woodwork is striking, since this is missing from European manuals until the end of the eighteenth century AD.

Besides Li's treatise, there was also *Ying Zao Zheng Shi* (Right Standards of Building Construction) of the Ming, while later dynasties issued more or less official technical compilations. But no work by any other individual ever took the place of the *Treatise on Architectural Methods*.

A quite different tradition of literature which is relevant is that of the descriptions, in prose or poetry, of cities, palaces and temples. From the Han onwards, rhapsodical odes on the successive capitals became a distinct literary genre, and in AD 530 a prince of the Liang brought together a great collection called the *Wen Xuan* (General Anthology of Prose and Verse). This contains poems about Xi'an and Luoyang, but the poetical phraseology is notoriously difficult, the technical terms obscure and the descriptions of buildings and layouts somewhat vague. All the same, they will repay study. There was also the *San Fu Huang Tu* (Description of Three Metropolitan Cities) by Miao Changyan, which may date from AD 140, though more probably from the third century. It contained drawings and charts and is considered a fairly reliable source for the later Han capital, Xi'an.

The tradition continued, and in the twelfth century AD onwards the victories of barbarians gave a powerful stimulus for the reminiscent description of cities and their buildings before the storms broke. Thus the *Dong Jing Meng Hua Lu* (Dreams of the Glories of the Eastern Capital), dealing with Kaifeng, appeared just twenty years after the city's fall in 1127. A century or two later, the Mongol invasion of the Song capital at Hangzhou gave rise to at least four books, among them the *Du Cheng Ji Sheng* (The Wonder of the Capital) and the *Meng Liang Lu* (The Past Seems a Dream). It would certainly be possible on reading such graphic accounts as these of the appearance and life of the great cities to derive much about their architecture and town planning.

A third great class of literature yields further opportunities of reconstructing the architecture and layouts of medieval Chinese cities and public buildings, namely that recording the work of archaeologists and local antiquarians. Nearly every city has its 'local gazetteer', a work on the history and topography of the place, often in many revised editions, and always including traditions of buildings and building plans. Most attention, of course, was given to the capitals. Typical was Wei Shu's *Liang Jing Xin Ji* (New Records of the Two Capitals) written during the first half of the eighth century AD. This laid a foundation for others, as did a map of Xi'an prepared in about AD 1075, which was drawn to a scale of 3 centimetres to 1 kilometre, identifying and marking the sites of ancient buildings.

Recent studies have shown, however, that any reconstruction of ancient and medieval buildings in the light of textual evidence only is extremely difficult. For interpreting the illustrative material from texts, whenever it can be obtained, is indispensable. Fortunately, we are not entirely without it. From the Warring States period, and from the Han and Jin dynasties, there are carved vessels, moulded bricks and tomb-models, all making a considerable contribution. For the Tang period, however, we are much better off, for we have the wealth of the fresco paintings of the Dunhuang cave-temples, many abounding in architectural detail. These may be divided into two groups: actual cities and temples of the period, and scenes in Buddhist paradises. These last invariably comprise halls, pavilions, galleries, forecourt, pools, platforms and bridges in elaborate ensembles. Everything in the paintings can be translated into accurate ground-plan and precise elevation.

A further source of information about the position of architectural science in the different periods would be the titles and powers of officials concerned. Thus the *Record of the Institutions of (the) Zhou (Dynasty)*, in its account of the Artisan-Carpenters and Engineers, says that they have to construct capital cities and buildings. Again a Directorate of Buildings appeared in the Northern Qi dynasty (AD 550 to 557). Before the Sui (AD 581 to 618), architectural officials were appointed only when there was need of palace or government buildings; at other times, the posts were left vacant. Gradually, changes of meaning occurred, and apparently, in the Yuan period (AD 1271 to 1368), the Directorate of Buildings was no longer a building department, but rather a part of the imperial workshops.

Principles of construction

In what has gone before we have been able to catch a glimpse of some of the essential principles of Chinese building. The moment has now come to penetrate further into the matter, and we shall find that certain obvious questions have illuminating answers, while to others there is still no satisfactory solution.

Of this second class is the problem of basic materials. Why was it that the Chinese throughout their history built in wood and tile, bamboo and plaster, never making use of the stone which in other civilisations, such as Greek, Indian and Egyptian, left such durable monuments behind? It certainly cannot be said that China had no stone suitable for great buildings, but it was used only for tomb construction, steles and monuments (in which typical woodwork details were frequently imitated), and pavements for roads, courts and paths. Possibly the answer lies in various cultural differences between China and the West. Social and economic conditions might throw light on the matter, for the forms of slavery known in China in different ages seem never to have paralleled those western usages which could despatch thousands of human beings at a time to hard labour in quarries. In Chinese civilisation there is absolutely no parallel to those great sculptured friezes of Assyria or Egypt which depict the harnessing of large numbers of workers in the transportation of enormous monoliths for carving or building. It might seem that no rule could have been more absolute than that of Qin Shi Huang Di, the builder of the first Great Wall, and of course there is no doubt that in ancient and medieval China very great labour forces could be mobilised by *corvée*, but what matters is the state of society in which the characteristic forms of Chinese architecture were originally determined, and it may well be that some connection exists between the timberwork style and the absence of mass slavery. On quite another plane, the ancient symbolic-correlation philosophy was perhaps also involved (this abridgement, volume 1, pp. 153–7). If stone was regarded as belonging to the element earth, it would have been proper for use only upon and under the ground, while wood was an element in itself, occupying a middle place between earth and the fiery essence of the heavens, hence the only fitting substance with which to build. But these notions are speculative, and the question remains.

Much more fruitful is that other question which so many foreigners in China must have asked themselves: what can be the nature and origin of the curving roof, that most characteristic and beautiful feature of Chinese buildings? The idea that it derives from an ancient desire to imitate the curves of tents and sheds of matting has been a popular cliché of tourists for centuries. No one, however, has found any authority for this, either literary or archaeological. Besides, it is the wrong kind of answer. What we really need to know first is by what constructional method the Chinese roof succeeds in getting its curve. This can be appreciated by examining Figs. 410 and 411.

As Fig. 410 shows, the great supporting columns rest upon plinths (*b*) which form part of the platform foundation (*a*), and they are fixed together

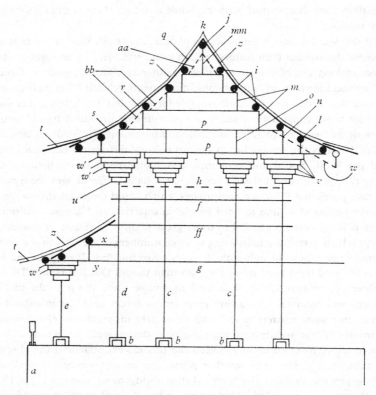

Fig. 410. Timberwork construction of a hall, from the *Ying Zao Fa Shi* (Treatise on Architectural Methods) of AD 1079.

a	platform *tai* [*thai*],[1] *jie, ji* [*chieh, chi*][2] (with balustrade *gou lan* [*kou lan*][3])	i	purlins *heng* [*hêng*],[15] *lin* [*lin*],[16] *lin tiao zi* [*lin thiao tzu*][17]
b	stone plinths for the wooden columns[a] *zhu, chu* [*chu, chhu*][4]	j	ridge-pole (uppermost purlin) *ji heng* [*chi hêng*];[18] *dong* [*tung*],[19] *fu* [*fu*][20] (can be purlins in general)
c	principal columns[b] *zhu* [*chu*],[5] *ying zhu* [*ying chu*],[6] *jin zhu* [*chin chu*][7]	k	roof ridge *zheng ji* [*chêng chi*][21]
d	external principal column supporting eaves[b] *lao yan zhu* [*lao yen chu*],[8] *yan zhu* [*yen chu*][9]	l	purlin supported from eave column *zheng xin heng* [*chêng hsin hêng*][22]
e	short external principal column supporting aisle eaves[b] *xiao yan zhu* [*hsiao yen chu*][10]	m	upper outer queen-posts (or blocks) *tuo dun* [*tho tun*][23] ('main beam stands')
f	main tie-beams[c] *da e fang* [*ta ê fang*][11] ('forehead beam')	mm	king-post *zhu ru zhu* [*chu ju chu*][24] (*zhuo* [*cho*]),[25,26] poetic word also applied to all vertical posts in the roof-timbering *shan zhu* [*shan chu*][27]
ff	other tie-beams *fang* [*fang*][12]		
g	lowest main tie-beam *kua kong fang* [*khua khung fang*][13] ('bestriding emptiness beam')	n	lowest outer queen-post *gua zhu* [*kua chu*],[28] *tong zhu* [*thung chu*][29]
		o	inner queen-posts *chen ke mu* [*chhen kho mu*][30] ('underclothes')
h	flat boards for coffer ceiling *ping ban fang* [*phing pan fang*][14]	p	main cross-beams[d] *jia liang* [*chia liang*][31] (no. 1, no. 2, etc.), also *liang* [*liang*][32]

q upper rafters *nao chuan* [*nao chhuan*][33] ('brain rafters'); all rafters may be called *chuan* [*chhuan*],[34] *jue* [*chueh*][35]

r middle rafters *hua jia chuan* [*hua chia chhuan*][36] (i.e. those corresponding to the layers of cross-beams)

s lower, or eave, rafters *yan chuan* [*yen chhuan*][37] ('eave rafters')

t cantilever eave rafters[e] *fei yan chuan* [*fei yen chhuan*],[38] *cui* [*tshui*][39] ('flying rafters')

u lowest corbel bracket unit *tou qiao dou gong* [*thou chhiao tou kung*][40] ('head of the tail-feathers')

v superimposed corbel bracket units (both parallel with, and at right angles to, the beams)[f] *dou gong* [*tou kung*][41]

w cantilever principal rafters *ang* [*ang*][42]

w′ false cantilever principal rafters *ang* [*ang*][42]

x extension beam supporting purlin of aisle roof *dan bu liang* [*tan pu liang*],[43] *ru fu* [*ju fu*][44]

y main aisle extension beam [*thao chien pao thou liang*],[45] *ru fu* [*ju fu*][44]

z tile surface alignment, with boarding *wang ban* ([*wang pan*][46]) beneath

aa inverted V-brace supporting the king-post (only in Song and pre-Song buildings) *cha shou* [*chha shou*][47] (derived from the ancient *ren zi gong* [*jen tzu kung*][48])

bb side braces connecting either cross-beam with cross-beam or cross-beam with purlin (only in Song and pre-Song buildings) *cha shou* [*chha shou*],[47] *tuo jiao* [*tho chiao*][49]

[a] Note the absence of a 'ground-sill' or tie-beam along the ground as in medieval European frame buildings. Yet the ground-sill is very prominent in China, not absenting itself from main doors and great gates. But it is concerned only with the woodwork planking of the curtain-walls.

[b] All these are connected together along the length of the building by tie-beams.

[c] Medieval 'bressumers'. Strictly, they should run directly under the lowest corbel brackets.

[d] Medieval 'somers'. The uppermost one was called the 'wind-beam'.

[e] Analogous to the 'firrings' of medieval European wood construction.

[f] The transverse corbel brackets (or bracket arms) are termed *hua gong* [*hua kung*].[50] The height of the standard *hua gong* selected is taken in the *Treatise on Architectural Methods*. The longitudinal corbel brackets, which cannot be shown in the diagram, have several names. The *gua zi* [*kua tzu*][51] is a longitudinal bracket arm midway between the wall and the eaves purlin; the *man gong* [*man kung*][52] is a longer one which is fixed above it so as to extend the support. The *ling gong* [*ling kung*][53] is a longitudinal bracket arm supporting the eaves purlin. The longitudinal tie-beams, also not seen in the diagram, are all termed *fang* [*fang*],[54] with qualifying adjectives.

[1] 臺	[2] 階基	[3] 鉤闌	[4] 柱礎	[5] 柱
[6] 檻柱	[7] 金柱	[8] 老檐柱	[9] 檐柱	[10] 小檐柱
[11] 大額枋	[12] 枋	[13] 跨空枋	[14] 平板枋	[15] 桁
[16] 檁	[17] 檁條子	[18] 脊桁	[19] 棟	[20] 桴
[21] 正脊	[22] 正心桁	[23] 柁墩	[24] 林儒柱	[25] 悅
[26] 椴	[27] 山柱	[28] 瓜柱	[29] 童柱	[30] 襯科木
[31] 架梁	[32] 梁	[33] 腦椽	[34] 椽	[35] 栭
[36] 花架椽	[37] 檐椽	[38] 飛檐椽	[39] 橑	[40] 頭翹科栱
[41] 科栱	[42] 昂	[43] 單步梁	[44] 乳栿	[45] 桃尖抱頭梁
[46] 望板	[47] 叉手	[48] 入字栱	[49] 托腳	[50] 華栱
[51] 瓜子	[52] 慢栱	[53] 偁栱	[54] 枋	

N.B. The sketch is purely schematic and takes no account of the sizes and strengths of the various component timbers. The proportions of the building as shown in the diagram are rather inelegant, but have been chosen for convenience of demonstration on the page of a book. At the same time, it is true that Chinese monumental buildings always tended to be much larger in length than in depth, and the cross-section here depicted is not unlike that of the famous hall of the Hōryūji temple in Japan, built in AD 670.

by massive tie-beams (*f*, *ff*, *g*) at various heights above the floor. Above them rise in tiers the main cross-beams (*p*), which carry the roof. These are upheld by many queen-posts (*m*) arranged at suitable positions, with usually only one central king-post (*mm*) at the top under the ridge-pole (*j*). Lengthwise, along the main cross-beams (*p*) and towards their ends, purlins (*i*) are placed; to these, the rafters (*q*, *r*, *s*) are fixed. A closely similar arrangement is used for aisles, galleries or verandahs, built in a lean-to manner.

The fundamental unit is thus a 'two-dimensional' framework capable of endless variation and adaptability; it is known as the method of frame construction. Based on the platform of rammed earth, it goes back in essence, as archaeologists have shown, to the Shang dynasty (second millennium BC). In ancient times, only the cross-beams (*p*) were used, but as time went on it was found that this placed excessive strain at the junction between columns and beams, failures tending to occur there. To avoid this, a number of corbel brackets (*u*, *v*) between the top of the column and the cross-beam were therefore introduced. Successively longer corbel brackets were then piled up on top of one another at the head of a column, so as to form what were essentially corbelled arches of wood supporting the cross-beam. The corbel brackets branched forth, not in one direction only, but in both, thus supporting both the lengthwise and the crosswise beams (see Figs. 411(*a*), (*b*) and 412). The *Treatise on Architectural Methods* has many illustrations of them (Fig. 413), showing how they were fitted with tenons and mortises, pegs and dowells (Fig. 414). But sometimes they are hidden by panelling.

The typical arrangement of the roof and subsidiary roof for an aisle or verandah is shown in Fig. 411(*c*), but during the Tang, and before, it was usual to support the eaves with their rafters and flying rafters (Fig. 410(*t*)) by means of purlins or brackets (Fig. 411(*d*)). There then developed another system, that of placing the purlins upon 'cantilever principal rafters' or

---

Fig. 411. Diagrams to explain Chinese and Western building construction. (*a*) The principle of corbel brackets (*dou gong*). (*b*) A more complex example. (*c*)Typical arrangement of roof and subsidiary aisle roof in a Chinese building. There are no principal rafters; longitudinal purlins are supported on the transverse frameworks in any desired profile. (*d*) Eave rafters and flying rafters on bracket-supported purlins. (*e*) Eave purlins supported from the ends of cantilever principal rafters or lever arms (*ang*). (*f*) Characteristic structure of a Chinese building. (*g*) Characteristic structure of a European building. (*h*) Corbelled half-arches of an English hammer-beam roof. (*i*) Inverted V-braces of early medieval Chinese buildings (*ren zi gong*). (*j*), (*k*), (*l*), (*m*), (*n*) Typical Chinese transverse frames of columns (*zhu*), tie-beams (*fang*) and cross-beams (*liang*). (*o*), (*p*) Comparison of the fundamental Chinese building design with that of Greek and Gothic building. Normally the Greek gable covered the perimetral colonnade. (*q*) 'Forked hand' struts forming a trapezoidal truss.

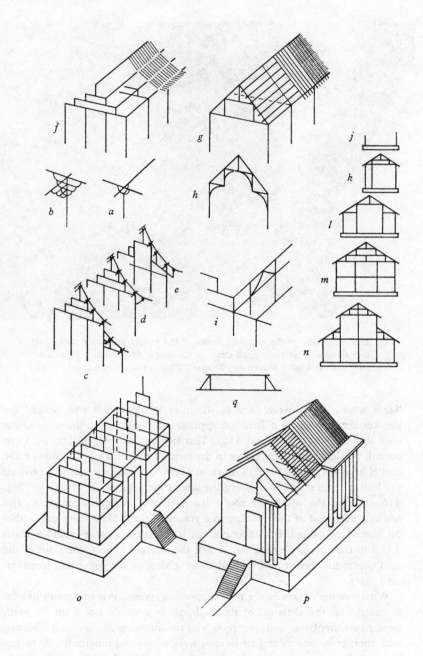

Fig. 412. One of the simplest forms of the corbel bracket system: part of the woodwork at the eighth century AD temple of Nanchan Si in the foothills of Wudai Mountain, Shanxi. (Photo. Joseph Needham, 1964.)

'lever arms'. These were fixed to the interior framework and pierced the bracket-arm clusters in a direction approximately parallel to the slope of the roof above (Figs. 411(*e*) and 415). This system died out during the Yuan period, but left permanent traces in the form of the 'bird's beak' ends on the corbel brackets (Fig. 410(*w'*)). In most regions, it was customary to raise up the four corners of the roof above the line of the edge of the main roof (Fig. 416), and in the south the roof edge itself was sometimes made to rise towards each end of the building in a graceful curve. Main roofs may either be hipped or gabled, but in important buildings the gables seldom reach down to the eaves; they are cut at half their height or less, the slant of the roof continuing below and around them, giving an imposing and harmonious effect.

Whatever the details of the roof supporting system, it is important to bear in mind that the skeleton of the building as a whole stood up by itself, needing neither base, walls nor roof. The two-dimensional horizontal beams, with their cross-beams and tie-beams, were connected longitudinally by tie-beams below as well as other tie-beams and purlins above, so that a three-dimensional construction came into being. This was a veritable ancestor of

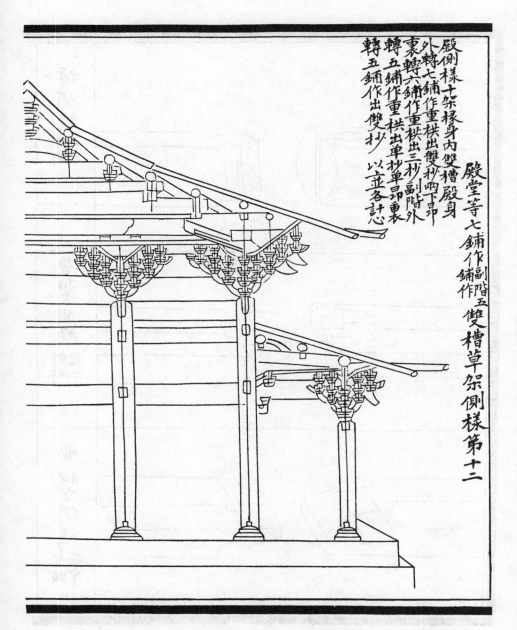

殿側樣十架椽身內雙槽殿身
外轉七鋪作重栱出雙杪兩下昂
裏轉六鋪作重栱出三杪副階外
轉五鋪作重栱出單杪即裏
轉五鋪作出雙杪
以並各計心

殿堂等七鋪作副階五
鋪作雙槽草架側樣第十二

Fig. 413. Diagram of three corbel bracket assemblies, from a cross-section
of a hall. The curving profile of purlins and the consequent curve of the
roof line is well seen. Lever arms are present. From the *Treatise on
Architectural Methods* of AD 1079.

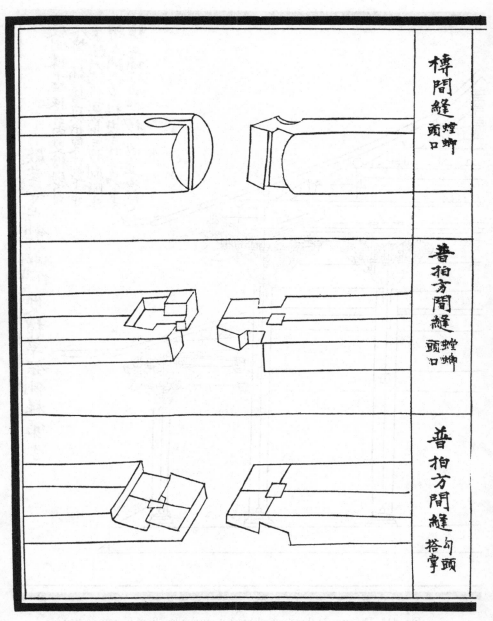

Fig. 414. Tenon and mortise work; forms of jointing in tie-beams and cross-beams. (Allowances must be made for distortions in copying.) From the *Treatise on Architectural Methods* of AD 1079.

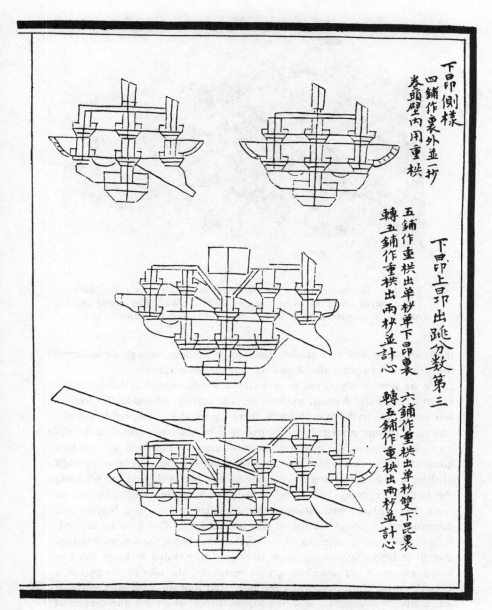

下昂側樣
四鋪作裏外並一抄
卷頭壁內用重栱

下昂上昂出跳分數第三

五鋪作重栱出單抄單下昂裏
轉五鋪作重栱出兩抄並計心

六鋪作重栱出單抄雙下昂裏
轉五鋪作重栱出兩抄並計心

Fig. 415. Bracket-arm clusters containing cantilevers; from the *Treatise on Architectural Methods* of AD 1079.

Fig. 416. The theatre-temple for New Year and other plays at the Ming fort of Jiayuguan at the western end of the Great Wall. The hipped gable roof curves gracefully up at all corners. (Photo. Joseph Needham, 1943.)

the steel-frame lattice of modern building technology, though far different from the sold casing walls of ancient and medieval Europe.

We are now in a position to understand another essential difference between Chinese and Western architecture. The curved roof and all that implies was impossible in the West because there the mind was wedded to straight and rigid sloping principal rafters, that is to say, a transverse slanting element. In China, on the contrary, the most important element was the longitudinal purlin, that horizontal element supporting common rafters; groups of these could be assembled according to any profile desired, by adjusting the framework itself. The corbelled half-arches of English hammer-beam roofs (Fig. 411(h)) were to some extent similar to the corbel bracket unit construction of China, but in no way liberated the West from its dependence on the sloping rafter. In China, on the other hand, the common rafters, though individually straight, were multiple; they ended at every third or fourth purlin, if not more often, thus permitting the tiles to descend in a smooth curve. Curiously, when the European roof did take on a curve, it was convex, not concave, as in the Mansard roof, where the queen-posts of the trusses are placed very much towards the sides of the building. It will be seen that the Chinese roof never had any principal rafters at all, though it did have the cantilever 'principals' mentioned above (Fig. 411(e)).

Fig. 417. Weight-bearing V-braces at the Foguang Si, a Tang building of
AD 857. (Photo. from Liang Sicheng (1952).)

What was the origin of the 'cantilever principal rafters' or 'lever arms',
consisting as they did of an intrusive triangular shape into the characteristic
Chinese square transverse frames (Fig. 411(j)–(n))? To answer this, one
has to go back into history. In ancient times, Chinese building technique
had made considerable use of double slanting joists meeting at a point like
the European roof-truss (Fig. 411(i)). These inverted V-braces were mainly
used as an ornamental device, diversifying the building as seen from the
front. They must have been in use as early as the first century AD, because
they appear in the *Shi Ming* (Explanation of Names) dictionary of about AD
100 as 'slanting struts'. However, there was one employment in which they
were not just decorative but made a significant contribution to structural
stability, and that was when they were used between the uppermost cross-
beam and the ridge-pole instead of a king-post (Fig. 417). This system is
extremely rare in those Chinese buildings which still exist, for by the end of
the Tang period the king-post had universally superseded it. Nevertheless,
this did not mean that 'forked hand' struts disappeared entirely; they existed
for some time as strengthening for the king-post. Not only that, but they
could be repeated at each end of the lower cross-beams, sometimes connecting beam with beam, sometimes beam with purlin, thus forming a kind of
trapeze-shaped truss (Fig. 411(q)). But because of their multiple character

they never in any way dictated the profile of the roof curve, which continued to depend solely on the layout of beams and purlins.

One can now see how easily the idea of the 'cantilever principal rafters' could have arisen. They were nothing but an extension of the principle of the 'forked hand' struts to solve the problem of the overhanging eaves. Of course, it may well be that the ultimate reference for a curving roof is due to the fact that in ancient times the 'front' of a Chinese building was the longer side, instead of the shorter one as in Egypt, Greece and medieval Europe (Fig. 411(*o*) and (*p*)). It was therefore natural enough to use complicated transverse framework partitions, since they would not spoil the perspective.

Whatever we may now think of the 'tent-theory', it is clear that the upturned roof-edge in China had the practical effect of admitting the maximum amount of slanting winter sunlight and the minimum downpouring summer sunlight. It also reduced the height of the roof while keeping a steep pitch for the upper part and a wide span at the eaves; and it thus reduced wind pressure from the side. This property must have been very important in reducing movements about the bases of the columns which simply rested on their stone plinths and were not generally taken into the ground. Another practical effect of the curving concave roof may have been the way snow and rainwater shot well off the eaves into the courtyard and so away from the edge of the platform. But of course it must always have given the greatest aesthetic satisfaction; as the late nineteenth-century army doctor, John Lamprey, suggested, it was the imprint of a particular taste and genius upon a structure intrinsically capable of responding.

As all the diagrams have shown, and our remarks have underlined, no part of the weight of the roof or structural beams was taken by the walls of Chinese buildings. Complete freedom was thus assured for placing windows and doors, and for their construction in delicate woodwork and lattice. A building could be remodelled without any danger of collapse, and all openings in walls could be as large as desired; indeed, in the hot climate of the south, one whole side of the hall could be, and often was, left open.

The use of iron columns and beams goes much further back historically than is generally thought. We shall see some unexpected examples in Chinese bridges (p. 117), while in AD 950 a hall built at Canton contained twelve pillars of cast iron each 3.6 metres long. Iron tie-bars were used fairly frequently in late medieval and Renaissance vaulting, but the real breakthrough in the West was only made in 1797 by Charles Bage, who constructed a five-storey flax mill at Shrewsbury, still standing and in good repair in the 1970s. But its cast iron beams and cast iron columns were joined by brick arches, and though it formed the first multi-storey iron frame building, lateral stability was still provided by massive external walls. It was forty years before a three-dimensional iron lattice stood up by itself: the Crystal Palace near London. Cast iron, wrought iron and steel were all used in the first buildings of skyscraper type, such as the ten-storey Home Insurance Building in Chicago in 1884, and the complete skeleton building followed soon afterwards. Thus it became possible to replace the walls of buildings almost wholly by transparent sheets of glass. But probably very few of the architects and engineers who took part in this great movement realised that a definitive escape from weight-bearing walls had already been accomplished by their Chinese predecessors during the previous 2000 or 3000 years.

Drawings, models and calculations

Any history of Chinese building technology, however brief, should devote some attention to the records which exist of the preparatory work carried out by the builders and architects of old. Here we can give only a few examples from the surviving literary and graphical material.

In the fifth century AD, the *Shi Shuo Xin Yu* (New Discourses on the Talk of the Times), speaking of a complex of buildings erected by the emperor Wen of the Wei State (AD 220 to 226), says:

> The Ling Yun Tai had towers and temples most elaborately and ingeniously built. All the pieces of timber were first weighed, so that there was a perfect balance (between the sides of the buildings). This was why the high buildings showed no sign of leaning over or collapsing, though some of the storeys were quite lofty, and often shook and vibrated in strong winds. The emperor Ming (AD 227 to 239), on mounting some of the towers, was alarmed at (what he thought was) a dangerous situation, so he caused one of them to be supported by an additional large column. (Some time after this was done) the tower collapsed and was destroyed. People talking about this result said that it was due to getting the weight unbalanced.

Pilot projects were also used to evaluate costs. In about AD 1197, the Neo-Confucian philosopher Huang Gan found himself charged with rebuilding the walls of Anqing, a city of which he was governor. So he began by constructing a trial length, which enabled a fair estimate to be made of the expense in manpower and materials. After that he pushed on the work with all speed, and it was successfully completed before the Jurchen Jin armies could attack the city.

During the time of the Northern Dynasties, drawings and models also appear. In about AD 491,

Fig. 418. This shows a remarkable model of a fortified manor house of the Han period dated AD 76 and found in Guangdong. Towers at each corner and two pavilions on the central axis enclose two model buildings of two rooms each; these when taken out reveal figures engaged in various farming and domestic activities. (Photo from Anon. (1902?).)

Cui Yuanzu said to the emperor that his nephew Cui Shaoyu was coming to the capital, and that he was exceptionally skilled in coloured architectural drawings. Cui suggested that the emperor should order him to make a model of (new?) palace buildings, and retain him. But the emperor did not feel able to follow the suggestion and so, after making painted diagrams of the palaces, Cui Shaoyu went home.

One would give a great deal to have these late fifth-century drawings and models now, so that we could see how much progress had been made beyond the simple designs seen in tomb-models of the Warring States and Han periods (Fig. 418). Miniaturisation was evidently coming into use, and models are often heard of in the Song (tenth to thirteenth centuries AD). For example, the *Qing Yi Lu* (Records of the Unworldly and the Strange) of AD 950 tells us that:

Certain workers made for Sun Chengyu (a general of the tenth century AD) a small model of Li Shan (mountain), complete with streams, bridges, houses, pavilions and paths, made of a kind of cake mixed with camphor. Subsequently a model in wax was made.

This would have been one of those landscaping designs, such as must have been prepared for the summer palaces we can visit today. The same source tells also of a maker of architectural models who prepared some kind of grotto-pavilions at about the same time.

It was just at this period that miniature buildings were greatly employed as a sort of interior decoration. When the Japanese monk Jōjin visited the capital Kaifeng in AD 1072 he noted that the great hall of the chief abbey of the Chan school had a ceiling that was 'all set out with (model) treasure-halls'. Libraries especially were treated in this way, the bookcases being crowned with whole temples in miniature. We can gain an excellent idea of what Jōjin saw because there still exists at Datong in the north a splendid library at the Lower Huayan Si Temple built in AD 1038. Here, over a central doorway, is a magnificent model pavilion on a flying bridge of cantilever type (Fig. 419). Beautifully modelled buildings have also been used for holding devotional relics.

In the seventeenth century AD, the writer Jiang Shao-Shu noted that there were those specialised in architectural drawings and paintings. Though these were not plans, they doubtless helped clients and builders alike, and their style was called 'sharp-edge painting' to distinguish it from that of the vaguer forms of misty landscapes. Jiang also claimed that artists wishing to succeed in this field usually became skilled in building calculations. It was a form of art to which much significance was attached in China.

This is why the Song *Treatise on Architectural Methods* of AD 1097 is such a landmark. Indeed, the excellence of its constructional drawings raises an issue of some importance. Nothing so far mentioned constitutes what we would now call 'working drawings', but the shapes of component parts of frameworks are so clearly delineated (see Fig. 420) that we can at last almost speak of working drawings in a modern sense – perhaps for the first time in any civilisation. Engineers of our own time are often inclined to wonder why the technical drawings of ancient and medieval times were so extremely bad; what remains from the Hellenistic world is so distorted as to need much interpretation, and the machine drawings of the Arabs are notoriously obscure. The medieval builders of European cathedrals were no better draughtsmen; even Leonardo himself produced little that was clearer than sketches, brilliant though these sometimes were. We must face the fact that Euclidean geometry had no power to give Europe precedence over China in the appearance of good working drawings, at least in building construction. Indeed, the very reverse was true.

Of the computational work necessary before any building was even started, many traces remain in the work of the eleventh-century architect Li Jie. But before leaving this matter, it must be noted that the majority of Chinese technical terms in solid geometry were derived from the preoccupations of

Fig. 419. Miniature buildings as interior decoration; a flying bridge pavilion of the Sūtra Repository at the Lower Huayan Si Temple at Datong in Shanxi. The great richness of the corbel bracket clusters in this model, made in AD 1038, is noteworthy, as is also the presence of two rows of cantilevers in its roof. (Photo. Joseph Needham, 1964.)

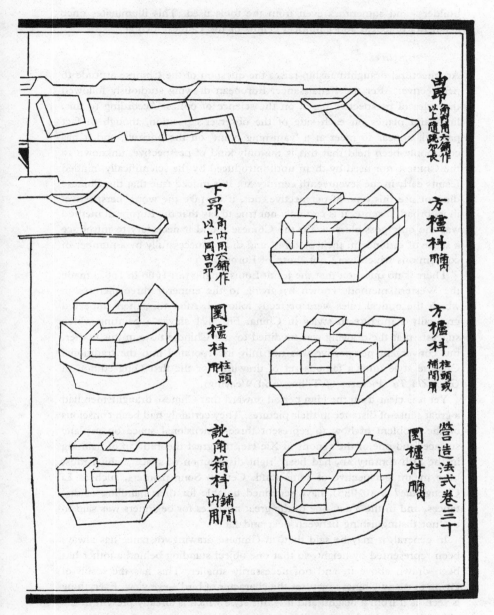

曲昂
角内用六鋪作
以上隨跳加長

下昂
角内用六鋪作
以上同由昂

方櫨枓
用

方櫨枓
柱頭或
補間用

圜櫨枓
柱頭

訛角箱枓
鋪間
内用

營造法式卷三十

圜櫨枓
鋪内

Fig. 420. Working drawings in the the printed edition of *Treatise on Architectural Methods* of AD 1079: five examples of bracket-arm bases and two cantilever arms.

builders, and sometimes even from the tools used. This illuminates once again the practical and empirical genius of the Chinese people.

Perspective

Architectural draughtmanship raises the question of the Chinese attitude to perspective. After the Renaissance, European drawing studiously followed the rules of perspective based on the science of optics. According to this, lines and planes on each side of the observer's position, though in fact parallel, appear to meet at a 'vanishing point' on the horizon, and it has commonly been held that this is the only kind of perspective, unknown to the Chinese nor used by them until introduced by the scientifically minded Jesuits early in the seventeenth century AD. It is indeed true that the Chinese did not use this type of perspective, but, if we take the word 'perspective' in its broadest sense, it is certainly not true to say that the European method was the only possible one; for the Chinese found it necessary to introduce a sense of distance in their pictures, and did so successfully by a number of conventions which were not those of Europe.

There is no question that the Jesuit Louis Buglio (AD 1606 to 1682) made the Western methods known by giving to the emperor three pictures in which the optical rules were perfectly followed. After this, there appeared gradually a mixture in styles in China. Fig. 421 shows a painting which suggests that the drawing was modified to suit Chinese taste in this matter, but convergent perspective was not fully incorporated into the traditional Chinese style until a famous set of drawings for the first Qing edition of *Geng Zhi Tu* (Pictures of Tilling And Weaving).

Yet it is clear that from the Han period onward that Chinese draughtsmen had a great sense of distance in their pictures. They certainly had been conscious of the problem of how to represent three-dimensional space on a plane surface, and one of the canons of Xie He, the great theoretician of painting in the fifth century AD, had been 'right distribution of space', which must have meant perspective of some kind. Certain Song painters, such as Li Cheng (*died c.* AD 985), have remained notable for their handling of distances, and in the Yuan one of the great mistakes for beginners was said to be 'not distinguishing between near and far'.

In general, it may be said that, in Chinese drawing, distance has always been represented by height, so that one object standing behind another has been drawn above it, and not necessarily smaller. This has the result of giving to many Chinese pictures the character of bird's-eye view. Everything is seen as if from a height, and it is this style which is already present in the oldest Chinese landscapes still in existence (first century BC). A curious consequence of it is that while, in a European painting, the spectator feels that he has the scene thoroughly under control, with the Chinese style the ground

Fig. 421. An example of mixed perspective principles in an eighteenth century AD scroll-painting of the Madonna; half European and half Chinese in character, this is probably the work of either the Jesuits Joseph Castiglione or Jean Denis Attiret, painters at the court of Qianlong; or of a Chinese artist influenced by them. The colonnade has an obvious vanishing-point, but the interior of the house follows the Chinese parallel (axonometric) projection. (Photo. from B. Laufer (1910).)

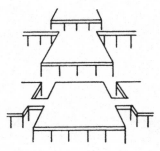

Fig. 422. A form of perspective seen in the Qianfodong (Dunhuang) frescoes – a series of superimposed vanishing-points.

surface starts from the distance and slips past under the spectator's feet to a goal infinitely beyond, that is below and perhaps behind the spectator (Fig. 422). In some cases this produces a feeling of uncertainty, of falling into the scene. Sometimes there even seem to be a series of plane surfaces, each with its own vanishing point, though this is rather unusual.

Let it be accepted, then, that on the whole there is no true vanishing-point in Chinese drawing, and no exact rules of foreshortening. The horizon boundary was not felt to be important; spectators were not compelled to participate in the drawing by their physical position. How then was it possible for the Chinese to delineate, as they did, the 'sharp-edge' quality of buildings? The answer is that they employed what we may term 'parallel perspective', that is a system in which lines which were parallel in fact remained so in the drawing. Fig. 395 (p. 29), for instance, shows how this works out. Reduced to its simplest elements, it can be appreciated by comparing it with the same drawing in convergent or optical perspective (Fig. 423). However, the Chinese convention could never show more than three surfaces of the interior of a room (that is, three sides only) or, to be more general, of a parallelepiped, whereas the post-Renaissance Western convention could show five (floor, ceiling and three sides). That the Chinese never attempted to solve the problem of five surfaces is simply the result of the fact that they lacked geometrical optics. There remains a further paradox, namely that the projection they did adopt is closely similar to those which architects and engineers use today for mechanical or working drawings. It is therefore unlikely that the absence of convergent perspective was a limiting factor at any time in China during what were the early stages of mechanical invention.

Parallel perspective can be found already in the drawing of the scenes carved in relief in the stone tomb-shrines of the Han period. Diagonal lines strike off from the front line of the picture, with figures or buildings along them. The convention continued in the works of the famous painter of the

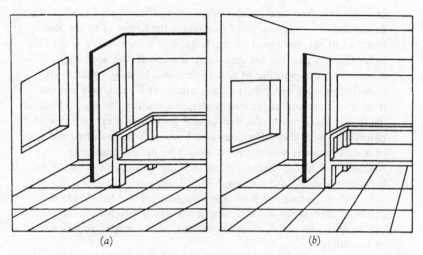

(a)   (b)

Fig. 423. Diagrams to show the contrast between (a) the Chinese parallel (axonometric) perspective drawing and (b) convergent or optical perspective.

fourth century AD, Gu Kaizhi, but with subtle modifications. Thereafter, it was never relinquished.

It may perhaps be said that parallel perspective, with its 'hovering view' region and the representation of scene depth by height, was an indication of an attitude towards Nature at once humbler and more social than that of Western man. Certainly, Chinese parallel perspective represents distance, but it is not founded upon the idea that the spectator is more distant from one part than from another.

It is fortunate that we have a striking defence of the diffuse view-region principle written in the Sung period, and paradoxically by the statesman-scientist Shen Gua, whom we so often quote. Li Cheng made some experiments towards a kind of optical perspective, as did Zhang Zeduan later, which is one of the reasons why his *Qingming Shang He Tu* (*Going up the River at the Spring Festival*), painted c. AD 1120 (Fig. 450, p. 135) makes such an immediate appeal to the modern eye. It was in about AD 1080 that Shen Gua wrote:

> Then there was Li Cheng, who when he depicted pavilions and lodges amidst mountains, storeyed buildings, pagodas and the like, always used to paint the eaves as seen from below. His idea was that 'one should look upwards from underneath, just as a man standing on level ground and looking up at the eaves of a pagoda

can see its rafters and its cantilever eave rafters'. This is all wrong. In general the proper way of painting a landscape is to see the small from the viewpoint of the large, just as one looks at artificial mountains in gardens [as one walks about]. If one applies [Li's method] to the painting of real mountains, looking up at them from below, one can only see one profile at a time, and not the wealth of their multitudinous slopes and profiles, to say nothing of all that is going on in the valleys and gorges, and in the lanes and courtyards with their dwellings and houses. If we stand to the east of a mountain its western parts would be on the vanishing boundary of far-off distance, and vice versa. Surely this could not be called a successful painting? Mr Li did not understand the principle of 'seeing tall from the viewpoint of the large'. He was certainly marvellous at diminishing accurately heights and distances, but should one attach such importance to the angles and corners of buildings?

Thus, the small viewpoint observed by a stationary individual was condemned in favour of the large panoramic view-area from which artists, embodying in themselves the visual experiences of a whole troupe of observers, could attempt to convey in its totality. The mountains and their detail was small, but what was large was the painter's mind and vision. Such was the orthodox attitude in all Chinese aesthetics.

A further point of great interest is that the Chinese developed an 'informative' as well as a 'representational' kind of perspective drawing. At first sight, such drawings appear very erratic; certain lines which ought to converge actually diverge, so that one can speak paradoxically of an 'inverted or divergent perspective'. What seems to have been done, for instance, in *The Bedroom Scene* by Gu Kaizhi, a painter from the fourth century AD, was to swing round the fronts or sides of the bed so as to make them appear nearer, and thus give the spectator certain items of information which would not otherwise have been available. Similar examples of divergent perspective can be traced in less marked form on Han reliefs. They can be seen also in some geometrical frieze motifs in the Dunhuang cave-temple frescoes from the Northern Wei to the Song periods. Lastly, it is interesting to note that both parallel and divergent perspective radiated from China to many other parts of Asia, especially Tibet, and the countries of the South Seas, notably Java, and Sri Lanka.

Dr Needham conjectures that experimental psychology may be needed to look into this matter of perspective from a different angle than that of comparative art history. He suggests this because it may well be that there exist differences in distance perception between different peoples.

NOTES ON THE HISTORIC DEVELOPMENT OF BUILDING

Words and traditions

As would naturally be expected, the technique of building goes so far back in history that it is worthwhile to look at what has become embedded in the structure of the ideographic language itself. There are three main radicals which have to do with dwellings: *yan* [*yen*] (𠂆 (厂)), which must have originally depicted a lean-to shelter against a cliff; *xue* [*hsüeh*] (𥤥 (穴)), which was originally a drawing of a cave- or pit-dwelling in rock or loess (wind-blown deposits ranging from clay to sand); and *mian* [*mien*] (𠆢 (宀)), which is, frankly, a roof. From these three origins derive the greater number of characters representing houses or parts of houses. Thus from *yan*, the cliff-shelter, come such words as *ting* [*thing*] (courtyard) (庭), *yu* [*you*] (the space under the eaves) (序) and *ku* [*khu*] (carriage-shed, treasury, arsenal) (庫), and these are but a few examples. From *xue*, not so prolific, came *chuang* [*chhuang*] (window) (窗) and *dou* [*tou*] (drain) (竇). Many familiar words arose from *mian*, such as *gong* [*kung*], (palace hall) (宫), *shi* [*shih*] (a private house) (室) and *tang* [*thang*] (reception-hall) (堂). Moreover, derivatives of the *mian* radical radiated in due course into fields much wider than that of building. For instance, the word family, *jia* [*chia*] (家), shows clearly from the radical 豕 that all homesteads were originally farms, for the roof has a pig underneath it, and the word for peace, *an* (安) shows a roof with a woman underneath it 安. Technical terms for constructional parts, however, are nearly all obtained from the tree-like radical for wood.

Lastly, *zong* [*tsung*] (宗), ancestral, a word of such far-reaching importance in Chinese culture, represents 宗, the symbol for a sign or omen set up within a house; it refers therefore essentially to the ancestral shrine or temple. The usual term for temple, *miao* (廟), comes from the lean-to radical 廟, but as for the significance of the objects portrayed within it, Dr Needham suggests that they seem closely related to the character for 'early morning' and the court ceremony which took place at that time, so that the reference may be to early morning worship. *Qin* [*chhin*] (寢), the sleeping room, seems to have had a brush under the roof 寢, probably because at first sleeping rooms were also storerooms.

Ancient tradition among the Chinese as to their earliest dwellings was rather precise. In a famous passage, the *Li Ji* (Record of Rites) of Han times says:

> Formerly the ancient kings had no halls or houses. In winter they
> lived in caves which they had excavated, while in summer they
> lived in nests which they had framed.

Fig. 424. Beehive huts from Tang frescoes at Qianfodong.

There can be no doubt that these winter dwellings were really holes in the ground, for shallow circular pits some 1 to 1.2 metres deep and 2.7 to 4.6 metres in diameter have been found by archaeologists investigating the Neolithic black pottery Longshan culture. Some of the dugouts were much larger than this, but all were covered by thatched roofs. Frequently dugouts used as storage pits are beehive shaped (1.8 metres deep and 1.8 metres in diameter at the bottom, but only 0.6 to 0.9 metres at the top), and Dr Needham suspects that this age-old shape was perpetuated above ground for the poorest of the common people as late as the Tang, since one can see many low beehive shaped huts of reed or thatch painted in a lifelike manner in the Dunhuang frescoes (Fig. 424).

A full excavation of a late Neolithic village has been carried out at Banpocun near Xi'an. Impressive in extent and detail, it indicates a population denser than in Europe at the same time (about 2500 BC), and more like Egypt or Mesopotamia. The floors of the houses are mostly circular in outline and about 5 metres in diameter, though some are oblong; all are sunk 1 metre or more below ground, are surrounded by a wall about 0.5 metre high, and have a hearth in the centre. Poles on each side of this supported the upper part of the roof, which certainly had a central hole, and a row of poles outside the wall held up the rim. Besides these dwelling pits, such villages as this also had many pits used for storage.

The upper, Yang, part of the dwelling, with its hole, allowed smoke to escape, and light and rain to enter, by the same route as the human occupants took to climb in and out. The lower, Yin, part contained the hearth and a cistern to catch the rainwater. One of the household gods, Zhongliu, took his name from the 'central drip', and the term came to stand for the tank. So also the stylised rounded vault of the later ceilings were called *tian jing* [*thien ching*] (天井) or *tian chuang* [*thien chhuang*] (天窗), and while *chuang* itself meant a window, as we have seen, *chuang* (囱) came to mean a vent or a flue.

As for the seasonal alternation in types of dwelling, Dr Needham suggests that they may derive from separate contributions to Chinese cultural evolution. In any case, the 'nests' have left no remains, but it is possible that they were rough shelters or houses built on piles, taking advantage of jungle trees

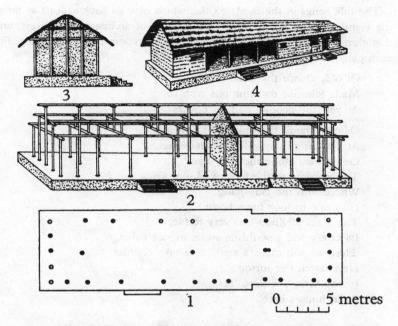

Fig. 425. Foundation of a Shang ceremonial building at Xiaotun, the area of the capital of Anyang, about 1250 BC. (1) The foundation of tamped earth (*terre pisé*) with steps still in position on one side, and all the pillar bases in their original positions; length 24 metres, width 8 metres, and height of platform 1 metre. (2) Reconstruction of the skeletal timber construction. (3) Probable appearance from one end. (4) Probable general appearance, suggesting a votive temple for ancestors.

as supports. If this were so, they could have given rise to the great tradition of the column based wooden structure of Chinese architecture. Indeed, it may even be that the raised harvesters' huts, which are so common a feature of the Chinese landscape at the appropriate season of the year, have come down to us almost unchanged from this remote antiquity. At all events, there can be no doubt that the type of building which was to become so universal in China, with its raised wooden pillars sitting on stone plinths above a platform of tamped earth, with walls later thrown round the structure, had already reached a highly developed state in the Shang period (second millennium BC). This is proved by excavations at Anyang, and Fig. 425 illustrates the plan and a reconstruction of one of these long buildings with thirty column bases. True to type, the main entrance seems to have been in the middle of one of the 24 metre-long sides, but the chief axis lay due north–south rather than east–west as became customary later.

The folk songs in the *Book of Odes*, which may go back at least as far as the eighth century BC, have several passages of architectural interest, and though not perhaps very informative, a few quotations are worth giving. For example:

> Of old, Danfu the duke
> Made kiln-like dwelling-pits with roofs,
> As yet (the people) had no houses.

> Of old, Danfu the duke
> At coming of day galloped his horses,
> Going west along the river bank
> Till he came to the foot of Mount Qi,
> Where with the lady Jiang
> He came to look for a home.
> The plain of Zhou was very fertile,
> Its celery and sow-thistle sweet as rice cakes,
> 'Here we will make a start; here take counsel,
> Here notch our tortoise'.*
> It says 'Stop', it says, 'Halt,
> Build houses here'.
>   . . .
> Then he summoned his Master of Works Si Kong[†]
> Then he summoned his Master of Lands Si Tu
> And charged them with the building of houses.
> Dead straight was the plumbline,
> The planks were lashed to hold (the earth);
> They made the Hall of Ancestors, very venerable.
> They tilted in the earth with a rattling,
> They pounded it with a dull thud,
> They beat the walls with a loud clang,
> They pared and chiselled them with a faint ping, ping;
> Three hundred rod-lengths[‡] all rose up,
> The drummers could not hold out.[¶]

Here the chief technical interest is the ramming of earth for the foundations and the walls of a temple. Another of these songs describes the erection of a feudal palace:

---

* This is a reference to use of the tortoise shell in a form of divination known as scapulimancy (see this abridgement, volume 1, pp. 191–4.)
† Note how old these titles are.
‡ Some 152 metres.
¶ The work was so enthusiastically carried out that they outdid those keeping the rhythm.

(The Lord) resembles and succeeds his forebears,
He builds a house of a hundred cubits‡
To the west and south are its doors,
There will he live and dwell, there laugh and talk.

They bind (the shuttering frames) one over the other,
They pound (the earth in them, it sounds) thak, thak;
This will keep out the wind and rain, the birds and the rats,
Here will the Lord be eaves-covered . . .

Again we have the moulds for the tamped earth, and a reference to the eaves.

Among other classical references to building and its traditions, we may notice one in the *Yi Jing* (Treatise on Arts and Games) of the third century AD, and another in the *Mo Zi* (Book of Master Mo) book of the fourth century BC. The Great Appendix in the former (perhaps of Warring States time, certainly Early Han) remarks on the change from primitive dwellings to buildings, while Mo Di, discussing the same subject, makes his usual attack on what he considers undue luxury and elaboration:

> Master Mo said: 'Before the art of building halls and houses was
> known, the people lingered among the hillsides and lived in caves
> or pit-dwellings where it was damp and injurious to health.
> Thereafter the sage-kings made halls and houses. The guiding
> principles for buildings were these, that (the house should be built)
> high enough to avoid damp and moisture, (that the walls should
> be) thick enough to keep out the wind and the cold, and (the
> roof) strong enough to stand snow, frost, rain and dew; lastly, that
> the partition walls in the palaces should be high enough to observe
> the proprieties of (separate accommodation for) the sexes. These
> things are sufficient, and any expense of money or labour which
> does not bring additional utility should not be permitted.

On the whole, it will be seen that these references do not give us much of information about the ancient technicalities. But a glance at *The Literary Expositor*, the most ancient dictionary (already mentioned), in its special section on buildings, will bring to light at once some twenty or so of the technical terms with which we have now become familiar (for example *fu* and *zhuo* [*cho*]); these it duly explains. There are also, of course, others which are rarer or which long ago became obsolete. Nevertheless, it remains clear from this that a substantial part of the technical vocabulary of building construction as we now know it was used by the architects of the Qin and Early Han (third century BC).

‡ Some 152 metres.

Periods and styles

We may accept, therefore, that the use of tamped earth platforms, halls with many wooden pillars standing on stone bases and suitable simple roofs was widespread from at least the third century BC onwards, while corbel brackets are characteristic of all Han buildings. Though none of these structures now survive, fortunately the Han people imitated their woodwork in pottery models as well as in the stone of tomb-chambers and funeral steles. These last were not mere slabs with inscriptions, but rather resembled 3.5 metre-high models of towers with elaborate roofs, the timberwork of which was faithfully reproduced in stone. Moreover, corbels in stone can be seen in a stone doorpost of the Han tomb of Feng Huan, dated AD 121 and found in Sichuan (Fig. 426). Again, in the Xiaotang Shan tombs of about AD 125, the capitals are drawn in such a way that indicates that the columns were topped by a series of successively longer brackets (Fig. 427). There were also moulded bricks which display much roof and other detail (Fig. 428); they also show that in the Han it is impossible to find any curvature in the roofs. Not until long after its close, perhaps not until the sixth century AD, does the discovery seem to have been made that the transverse frame permitted any desired variation from the straight line to be adopted.

A characteristic Han feature was the use of carved figures – draped females, or caryatides as we now call them – to form the columns. But there is little trace of it in later times, except that the columns of important buildings were (and still are) carved in high relief. Dragons, for instance, twine round them, as in the columns of the main hall at Jinci in Shanxi (Fig. 429).

During the period between the Han and the Sui, there was a gradual development of complexity, the corbel brackets becoming increasingly elaborate and the roofs more and more concave. By the earliest years of the Tang (AD 618 onwards), this was firmly established, for we find it not only in the oldest wooden buildings still existing in China, but also in still older structures preserved in Japan.

The chief new development in the period of partition often known as the Nan Bei Chao or Northern and Southern Dynasties (AD 479 to 581), and sometimes as the Liu Chao or Six Dynasties, was the construction of pagodas. As it eventually developed, the pagoda was a combination of the ancient Han towers of several storeys, known from tomb-models such as that shown in Fig. 430, and the *stūpa* forms of commemorative monument from India which ultimately imposed various curving silhouettes.

There was, of course, always an ancient Chinese inclination to build upwards. Zhou and Han tower building, somewhat overshadowed by the Buddhist 'spires' of later ages, gave the high raised platform of the Zhou,

Fig. 426. Corbels in stone can be seen in a stone doorpost of the Han tomb of Feng Huan, dated AD 121 and found in Sichuan. (Photo from Sickman and Soper (1956).) All such examples are modelled on gate watch-towers.

which was of pounded earth faced with brick, or possibly stone, and which sometimes carried a building above it. Such structures were used for diplomatic audiences, interviews between rulers, feasting, imprisonment, last ditch stands against enemies, as look-outs and not least for astronomical and meteorological observations. Indeed, the term Ling Tai or 'Numinous Tower' came to be the standard name for an observatory. Moreover, the ancient *Book of Odes* contains a celebration of the willingness of the people to build such a platform – the Altar of Heaven – for Wen Wang, the eleventh century BC founder of the Zhou dynasty. Then in the Han, Ban Gu's *Xi Du Fu* (Ode on the Western Capital (Chang'an)) of AD 87 mentions a tower built under Daoist influence by Han Wu Di to ensure communication with aerial spirits.

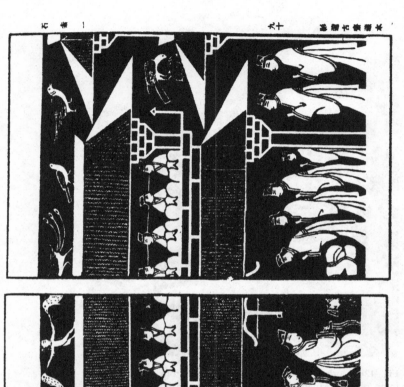

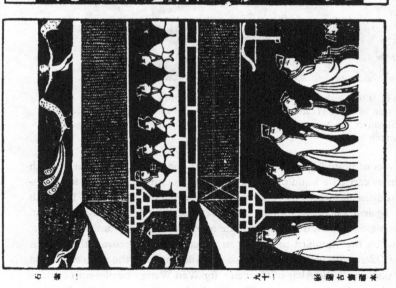

Fig. 427. A rubbing showing a two-storeyed hall depicted in the Xiaotang bas reliefs (c. AD 125). In the architecture of this Later Han reception hall, the clusters of corbel brackets at the top of each column are noteworthy, as also are the absence of any curvature of the roof lines. (Photo. from Anon. (1848).)

Fig. 428. A country manor house of the Han period depicted on a Sichuan moulded brick, probably intended for the decoration of a tomb-shrine. There is an entrance gate at lower left; to the right is the kitchen with well and stove; behind is a lookout tower with a ladder staircase, and in the courtyard there is a watchdog and a servant with a broom. At the back, on the left, the master is entertaining a guest while a couple of cranes dance in the garden. The corbel brackets at the top of the tower and the transverse framework of the reception hall are to be noted. The roofing of all the tamped earth walls is also characteristic. (Photo. from Anon. (1848).)

Fig. 429. Dragon columns of the main hall – the Hall of the Holy Mother – at the great Daoist temple at Jinci, south of Taiyuan in Shanxi. Built originally in AD 1030, its present form dates from the restoration of 1102. The front porch pillars lean markedly inwards, and the whole front sags in the middle, perhaps intentionally. In the elaborately painted structural woodwork, 'true' and 'false' horizontal beaked cantilevers alternate over the columns; three examples of each kind are to be seen in the photograph. This would be one of the earliest appearances of the false cantilevers. Inside the temple there is a remarkable set of around thirty wood and plaster approximately life-size statues, representing the attendants of the goddess and dating from the Song period. (Photo. Joseph Needham, 1964.)

Fig. 430. Origins of the pagoda; a Han pottery model of a tower. At each storey, corbel bracket woodwork supporting roofs and balconies can be seen. From a first or second century AD tomb at Wangdu in Hubei. (Photo. from Anon. (1849).)

Fig. 431. General view of the front of Buddha's Aureole Temple, standing among the misty foothills of Wutai Mountain. Dating from AD 857, it is the second oldest wooden structure still extant in China. The double and triple cantilever complexes are conspicuous among the plain assemblies. (Photo. Joseph Needham, 1964.)

Probably constructed of wood on a high stone-faced platform, it is said to have reached a height of some 114 metres. Though nothing of this kind has survived, the awe-inspiring formula can still be seen in the impressive guard-houses over the gates of Chinese cities. At all events, by AD 959, at the end of the Five Dynasties and Ten Kingdoms period, both curving roofs and pagoda towers had completed their basic development.

The Tang period (AD 618 to 907) seems to have been a time of architectural experimentation. The oldest extant wooden building comes from that time, and the next oldest, Buddha's Aureole Temple – the Foguang Si – which lies in the foothills of the Wuta mountain in Shanxi province, stands on a high platform adjusted to the slope of the mountain (Fig. 431). As the photograph shows, its massive columns, beams and corbel brackets have withstood the ravages of time since AD 857.

Next oldest to the Tang buildings is the great hall of the Protection-of-the-Nation Temple – the Zhengguo Si – at Pingyao in Shanxi, built in AD 963 under the short-lived Eastern Han dynasty. Following this is the more famous Joy-in-Solitude Temple – the Dule Si. The main gate and the God-dess of Mercy (Guanyin) Hall were built under Liao rule in AD 984. This is a very large work, a three-storeyed building housing a statue of the god-dess which is more than 18 metres high, and penetrating all three floor levels. More than a dozen different kinds of corbel bracket are used to suit the different positions (Fig. 432). A few other buildings also date from the early Song, particularly a great octagonal tower at Yingxian in Shanxi built in AD 1056 (Fig. 433). It has nine storeys, is more than 60 metres high, and has nearly sixty different types of corbel bracket.

Thus, from the tenth century AD, only half a dozen buildings have lasted, but there are at least fifty dating from before the fourteenth century AD which have been carefully investigated. Buildings of the Ming (fifteenth century AD) are comparatively common.

Somewhat naturally, attention has been concentrated on the oldest and most splendid of Chinese buildings, partly because their essential structure and the history of their development can best be brought out in this way, but something must be said of domestic architecture in China. This fulfilled its homely functions in a thousand beautiful forms, covering a land stretch-ing through thirty-five degrees of latitude and sixty of longitude.

In 1954, the English architect Francis Skinner made drawings of domestic buildings, a small selection of which are shown in Fig. 434. He recorded flat roofs of mud and wheat-straw used in the north from Gansu to Hubei com-bined with verandah and lattice windows, the stepped and shaped gables of Hunan, Jiangxi, and Guizhou, horned gables of shrines to the tutelary field-gods in Hubei, and the Cantonese farmhouse with its ridge terminals, central ornament and recessed bay entrance surmounted by a decorative carving. Convex barrel roofs somewhat like those of railway carriages occur in Liaoning, and in Gansu province the principle is extended to veritable barrel vaults of adobe brick that look like Nissen huts, probably due to being in a notoriously active earthquake region. The barrel vault is also found in the cave dwellings excavated in the loess hillsides of northern Shanxi and in the stone dwellings which the people build nearby in the image of them.

In the south-western provinces of Sichuan and Yunnan, the urban picture takes on an almost Spanish/Mexican character, in the sense that long ex-panses of blank white or grey wall capped only with coloured tiles alternate with highly ornamented and brightly coloured entrance gates and porticos. Rural manor houses in Sichuan stand out with their half timbering and white plastered walls against the groves of bamboos. Further eastwards, Anhui province, on the other hand, can boast a series of large courtyard farmhouses with exquisite interior carvings on beams and balconies; indeed, for the wealthier farmers and the scholar-gentry, the courtyard system lasted all through the ages.

Fig. 432. Tenth century AD woodwork at the Guanyin Hall of the Joy-in-Solitude Temple at Jixian in Hubei, built under the Liao rule. More than a dozen different kinds of bracket arms were used to suit the different positions in a lantern hall, so permitting the 18 metre image to penetrate all three floor levels. Several diagonal corbel brackets can be seen in the picture. (Photo. from Anon. (1954).)

Fig. 433. Octagonal wooden pagoda tower some 60 metres in height, at the Buddha Palace Temple at Yingxian in Shanxi. Built in AD 1056, it contains nearly sixty different kinds of bracket arms. (Photo. from Anon. (1954).)

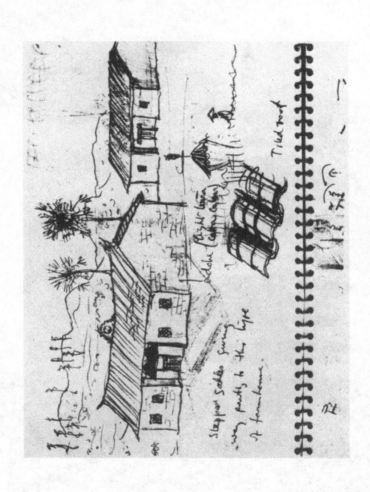

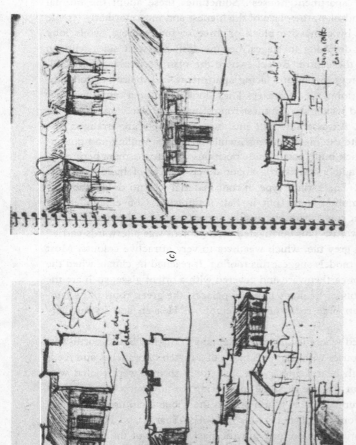

Fig. 434. Some leaves from a sketch book of the English architect Francis Skinner made in China in the 1950s. (a) shows free-standing farmhouses in Guangdong. These have light terracotta adobe brick walls, tiled roofs with central and end ridge ornaments, and recessed bays in the centre of the entrance side, having decorated doors and decorative carving above them. (b) Top and centre: homes with portico on the entrance side and prominent stepped gables. Bottom: typical house half-door entrance with lintel supported on shaped brackets; all in Henan. (c) Top: part of a row of houses with alternate blank and recessed bays, the latter with eaves supported on posts, and arched entrances flanked by openwork screens. The main transverse walls end in shaped gables. Centre: house with recessed bay forming portico. Bottom: stepped and decorated gable outlined with a white band. Though all in Henan, the latter is also typical of Sichuan and Yunnan.

The most extraordinary types of Chinese rural dwellings are those of the Hakka people of Fujian province. The need for security among an originally somewhat hostile indigenous population led to the development of fortified clan community 'apartment houses'. Sometimes these adopt the normal rectangular ground-plan, the place of the highest and most northerly temple hall being taken by a massive block of three or four storeys, while long wings, of height declining in stages, occupy east and west sides with an assembly hall in the centre. But elsewhere the plan is circular (Fig. 435), three or four storeys on inward facing apartments with balconies for individual families forming the periphery. These look down on a central circular courtyard, around which are set guest rooms, washing places and yards for pigs and poultry. An assembly hall and ancestral chapel are arranged diametrically opposite the main entrance, while lavatories, milling and pounding sheds, also brick-built, occupy side positions outside the main perimeter.

It remains to add a few words about other aspects of building in the different periods. The earliest type of roof material was no doubt the fully grown hollow bamboo stem split in half lengthwise. Convenient lengths were then laid in rows, with the inside curved surfaces alternately facing inwards and outwards. This corrugated arrangement was afterwards carried out in half-burnt grey tile, which weathers to very attractive colours. Most of the Han tomb models suggest this roofing. It reached its climax when the tiles were made of earthenware and covered with a bright glaze; such are the orange-yellow roofs of Beijing's imperial palaces, the green roofs of its temples and the deep blue roofs of the Temple of Heaven and its ancillary buildings.

Slates were used where locally available, and in the north and north-west there are many houses with flat roofs made of a thatch of branches and reeds surfaced over with beaten mud. The Han tomb-shrines were roofed with slabs of stone, but the use of stone in housing for any purpose other than the plinths of columns is now found only in the Tibetan culture-area and in a narrow zone along the Taihang scarp between Shanxi and Hebei.

Anciently, flooring was nothing but the packed earth of the foundation, and earthen floors continue in use in many parts of rural China. However, lime cement floors are favoured in the south, while floors of brick or stone are commoner in the north; both of these go back many centuries. Yet in large or important buildings, whether public or private, for hundreds of years the floors have been made of broad wooden planks.

Numerous accounts of travels in north China have familiarised the West with the use there of a simple form of central heating in domestic houses; this is a *kang* [*khang*], a raised built-in divan along one side of the room, made of sun-dried bricks or often simply tamped earth, under which a fire of any available fuel is stoked up from outside. What is not so generally

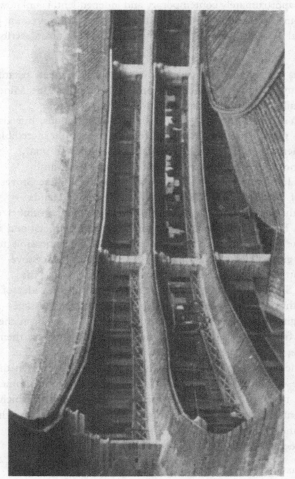

Fig. 435. Interior view of a four-storey apartment house of the Hakka people in Fujian, one of the most remarkable types of Chinese rural dwellings. The need for security among an originally somewhat hostile indigenous population is thought to have led to the development of these large circular communal buildings with walls mainly blank on the outside, defensible entrances, and public service facilities in the centre of the ring. This example is near Yongdin.

known is that the device was common from Han times, as shown in tomb-model houses. The Roman hypocaust for heating comes to mind, but the first mention of that appears relatively late (first century AD), though in the previous century the Roman engineer Vitruvius describes the essentials of it for heating baths. Whether either civilisation influenced the other, or whether the invention was approximately contemporary and independent, is unknown.

At any rate, Chinese central heating did not fail to make an impression on the Jesuits. Gabriel de Magalhaens, writing in about AD 1660, thus described it:

> This Coal is brought from certain Mountains two Leagues distant from the City (Beijing), and it is a wonderful thing that the Mine has never fail'd, notwithstanding that for above these four Thousand Years not only this City so large and populous, but also the greatest part of the Province, has consum'd such an incredible quantity, there not being any one Family, tho' never so poor, which has not a Stove heated with this Coal that lasts and preserves a Heat much more violent than Charcoal. These Stoves are made of Brick like a Bed or Couch three or four Hands Breadth high, and broader or narrower according to the number of the Family; Here they lie and sleep upon Matts or Carpets; and in the day time sit together, without which it would be impossible to endure the great Cold of the Climate. On the side of the Stove there is a little Oven wherein they put the Coal, of which the Flame, the Smoak and Heat spread themselves to all the sides of the Stove, through Pipes made on purpose, and have a passage forth through a little opening, and the Mouth of the Oven, in the which they bake their Victuals, heat their Wine, and prepare their Cha or The; for that they always drink their Drink hot . . . The Cooks of the Grandees and Mandarins, as also the Tradesmen that deal in Fire, as Smiths, Bakers, Dyers, and the like, both Summer and Winter make use of this Coal; the Heat and Smoak of which are so violent, that several Persons have been smother'd therewith; and sometimes it happens that the Stove takes Fire, and that all that are asleep upon it are burnt to Death. . . .

Every invention has its inconveniences.

All dwelling places have to be made comfortable with furniture. In the field of cabinet-making, China developed along unique and characteristic lines, which had a powerful influence on Europe in the eighteenth century. We cannot go into this subject here, except to note that already in about AD 1090 Huang Bosi wrote his *Yan Ji Tu* (Diagrams of Beijing Tables), and in AD 1617 Ge Shan wrote the *Die Ji Pu* (Discourse on Butterfly Tables), to mention only two works about furniture. Yet way back in the fourth century BC, lacquered tables, stools, beds and the like were in use, as princely tombs from the Chu State give evidence. But the primary unit of furniture was the low platform, which, in various shapes and sizes, served for kneeling or sleeping on, or as a table or arm-rest.

However, an interesting point arises in connection with chairs; a universal feature of all Chinese civilisation, as it was in Egypt, Greece and Rome. Yet throughout the intervening breadth of Asia (and in Japan) people squatted, knelt or reclined on the floor, with or without cushions. The answer to this seems to be that the chair was not known or used in China before the Early Han (202 BC to AD 9), and did not come into general acceptance before the end of the Later Han (AD 220), or possibly a little later. The camp-stool or folding chair cannot be established before the early third century AD, while the wooden frame chair did not find widespread use until the ninth century AD, near the end of the Tang. Dr Needham conjectures that there may have been a double evolution, the low wooden platform evolving in China into a set with back and arms, and the folding chair with a seat of cloth or leather arriving from somewhere in Central Asia. For in the time of Ling Di (AD 168 to 187) this was known as 'the barbarian bed' (*hu chuang* [*hu chhuang*]), and the name of the man who first produced it on a large scale (Jing Shi) has come down to us.

Pagodas, triumphal gateways and imperial tombs

The pagoda is a great feature of the Chinese landscape. The word landscape is chosen advisedly, since as we have seen its half-foreign origin from Indian Buddhism generally prevented it from arising within the city walls to compete with the drum-towers and gate-towers of cosmic-imperial authority. Its ancestor, the *stūpa*, an artificial mound, also had cosmic or microcosmic significance, since it was a model of the whole world, or at least the central sacred mountain, and it contained Buddhist relics at its heart, whence the superimposed parasols of honour from which perhaps the storeys of the pagoda ultimately derived. Its situation in or near Buddhist abbeys at some distance from the town gradually brought about a religious compromise with Daoist geomantic planning (see this abridgement, volume 1, p. 198), until the time came when no ring-walled city was complete without its pagoda standing nearby on the most suitable isolated hill to harmonise the earthly influences.

Such pagodas can have as many as a dozen storeys, with or without external galleries, and are sometimes square, sometimes many sided, but rarely circular; they may be of wood, more often of brick, but seldom of stone. They became, essentially, superimposed chapels, and were never intended as dwellings, even for monks. A particular type, the Tian-Ning style, so called from a famous monastery near Beijing, has a more or less unbroken tower from ground level to about one-third of its height, the galleries and storeys being repeated only above that level. Somewhat of this kind is the oldest extant pagoda, that on the sacred mountain of Song Shan in Henan. It is of brick and was built in AD 523 (Fig. 436) under the Northern Wei, and its spire represents a very sophisticated stage of development, for it must be assumed that earlier towers were much simpler.

Entirely true to its principle of building from repeatable single units or modules, Chinese culture contains many square one-storey buildings which represent, so to say, a pagoda's base or storey in isolation; the single cell apart from the body as a whole. As we shall see in the next chapter when discussing bridges, the arch, in the shape of the barrel vault, was known and used in China probably in the Zhou and certainly in the Han. But the true arch with its keystone at the top was not greatly used in the building of pagodas; the commoner construction was the vault with corbels to support the vaulting across the curving roof. These can readily be seen in small one-storey shrine buildings, such as the 'Alchemy Tower' (Liandan Lou) built on the forested hill behind the Daoist abbey near Zhouzhi south-west of Xi'an (Fig. 437). It consists of a single brickwork chamber entered by one door and roofed with a corbelled vault, the layers of bricks rising from the angles to form a series of corbels carried across the corners (Fig. 437 (c)). The brick was of a burnt red unlike anything now commonly seen in the neighbourhood, and was arranged in billet mouldings outside (Fig. 437 (b)). It was possibly of the Tang period, but because its eaves curl up at the corners, it may be early Song, and was probably originally used for alchemical experiments.

The ancient Indian *stūpa* was not completely absorbed into the pagoda; it continued on the ground throughout China, used as tombs or pious shrines, thus continuing one of its earliest functions. Hence the desert near Qianfodong is strewn with the exquisite shapes of these structures (Fig. 438) commemorating monks of the Song or Yuan.

As for the technical principles of pagoda building, depending as it did on wooden frameworks and the bonding of bricks or masonry, they were really only the extension of the building techniques to a particular specialised field. It may be guessed, however, that the simple truss forms represented by sloping struts proved especially useful in high wooden towers. There is a significant story about Yu Hao in this connection. Yu Hao, who, as the reader will remember (see p. 59), was the Master-Builder who constructed the Kaibao pagoda in Khaifeng in AD 989 as well as other famous buildings, and was also the author of the *Timberwork Manual*. In his *Dream Pool Essays*, Shen Gua has an entertaining story about the advice Yu Hao gave to another artisan-architect some ten years later.

Fig. 436. The fifteen-storey brick pagoda at the Temple of the Sacred Mountain of Song in Henan, built in AD 523 under the Northern Wei. Elegant with its twelve sides, it exemplifies the statement that for a pagoda one must pile Indian shrines on the top of a Chinese tower. (Photo. from Anon. (1954).)

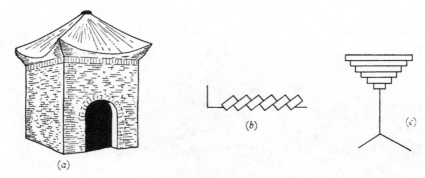

Fig. 437. The 'Alchemy Tower' at the Daoist abbey of Louguan Dai near Zhouzhi south-west of Xi'an, Shaanxi. (*a*) General view; (*b*) exterior brick billet moulding (in plan); and (*c*) interior corbelled vaulting carried across the corners and supporting an octagon, then a circle and finally a small square.

Fig. 438. Crumbling *stūpa* tombs in the desert across the dry river bed from the Qianfodong cave temples, some of which can also be seen in the escarpment of the background hills. On the left, the northern end of the oasis can be seen, and the tomb of Wang Dao-Shi, discoverer of the famous Dunhuang library, is nearby. These tombs, commemorating monks of the Tang, Song and Yuan periods, decay extremely slowly because of the dryness of the climate, though their coloured plasterwork is scoured and polished by the sandstorms of many ages. They extend over the desert for kilometres; each one has a different design, and every one is beautiful. (Photo. Joseph Needham, 1943, taken in the early morning.)

When Mr Qian (Weiyan) was Governor of the two Zhejiang provinces, he authorised the building of a wooden pagoda at the Brahma-Heaven Temple (Fantian Si) in Hangzhou with a design of twice three storeys. While it was under construction General Qian went up to the top and was worried because it swayed a little. But the Master-Builder explained that as the tiles had-not yet been put on, the upper part was still rather light, hence the effect. So then they put on all the tiles, but the sway continued as before. Being at a loss what to do, he privately sent his wife to see the wife of Yu Hao with a present of golden hairpins, and enquire about the cause of the motion. (Yu) Hao laughed and said: 'That's easy, just fit in struts to settle the work, fixed with (iron) nails, and it will not move any more.' The Master-Builder followed his advice, and the tower stood quite firm. This is because the nailed struts filled in and bound together (all the members) up and down so that the six planes (above and below, front and back, left and right) were mutually linked like the cage of the thorax. Although people might walk on the struts, the six planes grasped and supported each other, so naturally there could be no more motion. Everybody acknowledged the expertise thus shown.

Surely we have to deal here with slanting struts inserted in an otherwise purely rectangular network of diagonal wind bracing.

A remarkable department of pagoda building was that which made them of cast iron, or more often of bronze. These masterpieces have aroused the astonishment and admiration of foreign travellers in China from the Japanese monk Ennin in the ninth century AD to W. D. Bernard in the nineteenth century. The oldest existing iron pagoda, The Jade Springs Temple (Yuquan Si) at Dangyang in Hubei province, which dates from AD 1061, is of a very considerable size, being just over 21 metres high and having thirteen storeys. Its weight is about 54 tonnes. Another smaller one, of nine storeys, is the Sweet Dew Temple (Ganlu Si) at Zhenjiang in Jiangsu; this is probably the work of Pei Qu (*flourished* AD 1078 to 1086). In other cases, a masonry core may be clothed with cast iron plates, as at Beiducun, north-west of Xi'an.

Another gift from Indian to Chinese architecture was the triumphal gateway or *pailou* [*phai-lou*]. This is a free-standing gate of wood or stone, with superimposed lintel beams, erected for commemorative or triumphal purposes on an approach to a tomb, temple or palace, or even across any village road or path. Its name indicates that it was to carry aloft a notice, often an epigram. The traveller on the stone pathways of Sichuan comes upon relatively simple ones from time to time, proclaiming the name of a virtuous widow or a popular magistrate. Greater occasions call for three, five or seven

Fig. 439. A stone triumphal gateway in one of the streets of Qufou in
Shandong, site of the tomb and Temple of Confucius. (Photo. from
Forman & Forman (1960).) In somewhat simpler forms, this kind of stone
gateway is widely found, imitating woodwork in its roofs and bracket arms.

arches in a row (Fig. 439). Thus, in the seventeenth century the Jesuit
missionary, Louis Lecomte, wrote:

> The town (Ningbo) is still full of monuments called by the
> Chinese Paifam (*paifang* [*phai-fang*]) or Pailou, and by us
> Triumphal Arches, which are very frequent in China.
>
> They consist of three great Arches Abreast, built with long
> Marble Stones. . . . they adorn it with Inscriptions, figures and
> Embossed Sculptures of a wonderful beauty, with Knots wrought
> loose one within another, with Flowers curiously carved, and Birds
> flying as it were from the Stone, which in my Mind are
> Masterpieces.

Fig. 440. Ming imperial tombs; the tomb-temple of the Yongle emperor (*reigned* AD 1404 to 1424). It stands behind the main Hall of Sacrifice and in front of the broad tumulus containing the burial chambers surrounded by a rampart. It is known as the Brilliant Tower, and it shelters a large stele mounted on a stone tortoise. The human figures on the wall give an indication of the size. The object in the middle foreground is an open-air altar. (Photo. Joseph Needham, 1952.)

Imperial tombs have been said to be one of the great forms of Chinese architectural achievement. This is because the whole pattern they constitute is perhaps the greatest example of the co-option of wide tracts of landscape as part of the architectonic whole. Today they are seen as a monument to the architects and building workers who designed and made them, as well as to the emperors whose lot it was to order them for their burial. From the Han to the Sui dynasties (221 BC to AD 618), there remain nothing but tumuli, and from Tang and later there are only tumuli and some incomplete lines of battered statues. In a number of places, such as Mukden in Manchuria, there are tomb-temples of the early Qing emperors which are perfectly preserved, but the greatest masterpiece is that complex of Ming tombs in the mountains north of Beijing, and now known as the 'Thirteen Tombs'.

These cover a wide valley in the hills, each battlemented tumulus usually lying on the slope of a spur between two side valleys. Each is fronted by halls and temples in a vast compound populated by trees. Advancing along a road from the capital, the visitor meets first a splendid *pailou* of five arches (AD 1540) and then a solid triumphal gate-tower with three barrel-vaulted tunnels. After a distant vista of misty mountains, there comes a colossal pavilion open to the weather on each side through four great arches, and containing the largest inscribed stele in China, poised on a stone tortoise (AD 1420). This building is guarded by four ceremonial columns carrying stylised clouds. Then, as the path curves slowly to the right, there are stone statues on each side representing camels, elephants, horses and mythical animals, as well as civil and military officials; this ends in another *pailou*. After crossing bridges, the majestic roofs of the temples can be seen. Along a serpentine way paved with great flat stones the tomb of the Yongle emperor (AD 1424), the greatest of the family, is reached (Fig. 440). This is surrounded by a wall and gatehouses. A pavilion housing steles records the first Qing emperor's instructions to the local city magistrate to maintain in perpetuity these monuments of a conquered dynasty. After further courts and an altar is the Spirit Tower, carrying a large pavilion containing another stele, while in front is the main ancestral hall, by twenty-four giant cedar columns, each over 3.5 metres in circumference and 18 metres high. From the tower, a magnificent view of the whole valley can be seen, giving a sublime sense of this vast organic plan of landscape and buildings.

4

Bridges

When the architect Frontinus, writing about the aqueducts of Rome in the first century AD, completed his description, he added the following: 'With such an array of indispensable structures, carrying so many waters, compare, if you will, the idle pyramids, or the useless, though famous, works of the Greeks.' His Chinese counterparts would have had some sympathy for the attitude of mind which lay behind this remark, for no small part of their civilisation lay in a subtle combination of the rational with the romantic. This had its consequence in their structural engineering. No Chinese bridge lacked beauty, and in many this was present to a remarkable degree.

In describing the achievements of Chinese bridge building, it will be best to follow a logical classification, beginning with what is presumably the simplest type of bridge – a beam of wood or any rigid material laid straight across a stream or other obstacle to be spanned. Here, limiting factors are soon reached, for, before any considerable length is reached, the material will cease to support the weights which its builders will want to send over it. As we shall see, the Chinese explored the possibilities of this simple method up to the maximum strength of the strongest natural material available, by constructing a series of notable megalithic bridges.

Release from the narrow limitations of single beams or blocks came only with the development of the truss, in which many component members, each being only under tension or compression, are jointed together in a net-like geometrical system. Such beam bridges were fully developed only during the Renaissance in Europe, possibly due to discovery arising from a study of timber 'centering' which had been used (in China also) for the construction of arches. From geometry, the Renaissance engineers knew that the triangle was the only figure which could not be deformed or distorted without changing the length of at least one of its sides; hence their elaborate combinations of triangles in the trusses they built. Drawbridges must also come under the heading of beam bridges, and we shall have to say something about the different kinds of piers employed when the bridge has more than one opening. These may be wooden piles, or occasionally wooden tripods, stone in various designs and, last but not least, boats to make those floating bridges or pontoons which appear very early in history.

The next class of bridges is that in which the cantilever beam is used. Such a beam is one rigidly fixed at one end and free at the other, so that it can move slightly according to its flexibility. In cantilever bridges, a series of such beams are thrown out from both sides of the gap and connected by a beam or truss in the centre. The home of these bridges seems to be the Himalayan region, and they were early known and used in China.

Arches have formed perhaps the most frequent and widespread type of bridge. They were originally semicircular, and long remained so. In Europe there was a persistent belief that the Roman and Norman semicircular arch was indispensable because it directed the line of thrust vertically downward at the piers or abutments. This theory was not affected by the use of pointed arches in the Middle Ages, which have lines of two much larger circles or other curves crossing at the crown. The great departure from precedent came in the fourteenth century AD, when the base diameter of the semicircle was allowed to sink, as it were, far below the river, and the bridge therefore became segmental, leaping forth from its abutments like the flying galloping motif in art. For such bridges the abutments had to be made stronger. Here we shall see that this fundamental advance had been anticipated by a Chinese engineer of genius some seven centuries before its appearance in Europe. Other shapes of curves, such as ellipses, could, of course, be used, and were.

The last important class is the suspension bridge. Here the support comes from above by ropes or chains hanging from two stays and looping down in a 'catenary' shaped curve. In all the more primitive forms, the passengers and animals followed the curve as they crossed, but, perhaps as a development of handrails connected at short intervals with the deck of the bridge, there gradually arose the true flat-deck suspension bridge. The suspension bridge is native to many parts of the world, both Old and New, but only in one of them did the early engineers make the transition from supporting ropes to iron chains. This occurred at an early date in the very mountainous country of south-west China bordering on Tibet, Assam and Burma. The later suspension bridges of Europe derive from this remarkable iron-chain development.

We may now summarise the classification, inserting some figures for the spans which traditional Chinese engineers achieved.

Type of bridge	Construction		Maximum span (metres)
Beam	iron		3
	wood		6
	stone		21
Cantilever	wood		40
Arch	stone	semicircular	27
		pointed or two-centred ('Gothic')	21
		segmental	61
Suspension	catenary	single rope	
		V-section rope	
		tubular rope network	
		decked bamboo rope	137
		iron chain	
	flat deck	bamboo rope	
		iron chain	

It is of interest that the existence of different types of bridges seems to be betrayed in the structure of certain characters. It has been suggested that the earliest graph for *liang* (see inset), 氼, 朵朵 which means a beam or a bridge, was a drawing of a plank across a stream. There is no early version of the present character for *liang* (梁), but a related form contains the components of water and rice, while a third which seems to have been originally a drawing of a man doing something. Perhaps he was building an irrigation dam across which one could walk. In the early *Book of Odes*, the character is usually used to mean a fixed fish trap made of palisades, which could very naturally have formed a bridge. The most usual term for bridge, *qiao* [*chhaio*] (橋), adds the radical for wood to form *qiao* [*chhaio*] (喬), which meant high and arched, as evidently appears from the graphs of ancient times (朵朵, above).

Foreign admirers of Chinese bridges could be adduced from nearly every century of the empire. Between AD 838 and 847, the Japanese monk Ennin never found a bridge out of commission, and, when on his way from Shandong to Chang'an, he marvelled at the effective crossing of one of the branches of the Yellow River by a floating bridge almost 305 metres long, followed immediately by a bridge of many arches. In the last decades of the thirteenth century AD, Marco Polo reacted in a similar way, and speaks at length of the bridges in China, though he never mentions one in any other part of the world. The 12 000 bridges of Hangzhou, famous as an exaggeration of 'Marco Millione', probably arose from the omission of a line of manuscript and a confusion between city-gates and bridges; in fact there were in his time exactly 347, of which no less than 117 were within the city walls.

The first Renaissance visitors to China also conceived a great admiration for the bridges which they found there. One of the earlier of them, Galeote Pereira, wrote in about AD 1577 of the megalithic type, 'As you come into either of these cities (near Zhangzhou in Fujian) there standeth so great and mightie a bridge that the lyke thereof I have never seen in Portugall nor els where. I heard one of my fellows say, that he told in one bridge, 40 arches.' Parallel passages are frequent among most of the writers of the sixteenth and seventeenth centuries AD. Indeed, when Peter the Great sent an embassy to China in May 1675, one of the requests made to the Chinese by the envoy Nikolaie Milescu Spătarul was that expert bridge-builders should be sent westwards to teach their methods to the Russians.

It is interesting that one of the things which the early Portuguese visitors to China in the sixteenth century found most extraordinary about the bridges was that they existed along roads often far from human habitations. 'What is to be wondered at', wrote Gaspar da Cruz, the Dominican who was there in AD 1556, 'is that there are many bridges in uninhabited places throughout the country, and these are not less well built nor less costly than those which are nigh the cities, but rather they are all costly and very well wrought.' In such ways did the works of an all pervading imperial bureaucracy impress the visitors from an essentially city-state civilisation.

BEAM BRIDGES

The simple wooden beam bridge with trestle piers is found in most parts of China. Its chief interest is that it seems to have persisted unchanged from high antiquity. The famous scene of the 'Battle on the Bridge' in the Wu Liang tomb-shrine carvings, dating from about AD 150, shows clearly the graded approaches and the central span. It was probably this kind of bridge that the philosopher Mencius was speaking of in the fourth century BC, when he mentioned the seasonal repairing of footbridges and carriage bridges. Spans could not well exceed 4.6 to 6 metres between the trestle piers, but over shallow water the spans might be numerous.

This was necessarily the case when rivers of considerable width had to be bridged, as they were, very successfully, from quite early times. After the rise of the Guannei region – the 'area with passes', which includes Xi'an and Luoyang – as the centre of Chinese culture, the crossing of the Wei River became particularly important. A beam structure of many spans, the Heng Qiao, was built by Prince Zhao Xiang of Qin soon after his accession in 305 BC. Linking the capital of Xianyang with the lands and passes south of the river, it retained all its importance during the Han, when Chang'an on the southern side became the capital. Since its length, some 610 metres, and the number of spans, 68, are known, they must have each been approximately 8.8 metres long. While all its beams were of wood, giving a deck width of 8 metres, its piers were of stone in its northern section. We can visualise this rude but noble bridge in two ways: first by studying representations on Han bricks still in the museums of Sichuan (Fig. 441), as old as the Wu Liang relief, if not somewhat older; and secondly by looking at structures of the same type still existing today. Three rivers fall into the Wei near Chang'an, the Ba and the Chan to the east and the Feng to the west, and their old bridges of Han type with up to sixty-seven spans remain still as Ennin saw them when he walked across in AD 840 (Fig. 442). All such bridges are close to the water-level and disappear from sight at flood seasons.

Throughout Chinese history this style persisted. We hear of a trestle structure built by Zhang Zhongyan in AD 1158 ten *li* (about 5 kilometres) in length. In pictures by Song painters, such as Xia Gui, we find elegant bridges of wooden beams, with pavilions crowning their central spans (Fig. 443).

Besides wooden trestles, wooden piles or stone piers, all kinds of other supports for beam bridges have been used in China at one time or another. In the neighbourhood of Yazhou it has long been customary to rest bridges of wooden beams crossing the Qingyi River and its tributaries on 'gabions' made of open-work bamboo baskets filled with stones. One can even find instances of the use of iron columns as bridge piers. Some time during the Song, a Jiangxi man, Zang Hong, built a bridge based on twelve of these iron columns at Fouliang Xian in Sichuan, but they were replaced by stone ones towards the end of the dynasty. Yet five more bridges of this kind have been discovered in the literature, one in Yunnan, having seven spans averaging nearly 14 metres set on 12 metre columns, presumably composite. For four of the six, cast iron is distinctly stated, and it was probably used in all.

Even for this simplest type of bridge, effective pile-driving is needed, and until very recent times the Chinese continued to use an apparatus similar to that adopted by Vitruvius in about 30 BC and shown in Fig. 444 on the Old Silk Road. In this the hammer is raised to the top of its travel by a cable forking into many pulling ropes so that the combined effort of a number of men may be used. For smaller jobs, four to eight men operated a 'rammer' – a cylindrical stone with bamboo handles – while they stood on a small platform attached near the top of the pile, so that their weight added to the blows.

Stone beam bridges are familiar to the English because of the small 'clapper' bridges of the West Country. But in China the principle was used on a much larger scale; even the standard type of small stone bridge (Fig. 445) is considerably larger than any English clapper. When the span is less than some 5 metres and the height above high water 1.8 to 3 metres, this is the

Fig. 441. Chariot and horsemen crossing a pile-and-beam bridge with balustrade; a scene on a moulded brick of Han date from Chengdu, Sichuan. (Photo. Anon. (1848).)

Fig. 442. The Feng River bridge near Xi'an to the north-west; it is a pier-and-beam structure of Han type. Each pier consists of three pairs of pillars built of stone cylinders like threshing rollers and based on three disc-shaped roller-mill baseplates (hidden in the photograph by concrete added later). These in turn were supported on cypress piling. (Photo. Joseph Needham, 1964.)

type most commonly used. Such bridges are always constructed in a dry enclosure formed by a 'cofferdam', which is made by using a double row of bamboo poles tied together and covered with matting, the intervening spaces being filled with clay. Water is then pumped out using a square-pallet chain-pump (see p. 263) until the dry bed of the stream can be worked. Such cofferdams are made convex to resist the main water pressure of the river.

Three chief types of pier are used with these bridges. In the first and simplest form, long flat stones, like planks, are set up vertically parallel to the direction of the water flow, and mortised into a long foundation stone beneath. At the top, they are assembled and mortised into another long horizontal stone, longer than the width of the bridge which it supports. This system offers little resistance to the current, and can well sustain the impact of boats. The abutments look massive but are made with great economy, being no more than thin retaining walls with a filling of clay and stone chips. A second type of pier is also constructed of stone but in the form of a

Fig. 443. A wooden beam bridge, the centre span of which is topped with a pavilion. This type of structure, humped slightly to make easier the passage of craft under the central span, was perennial through the ages in China. The present example is from a painting by the eminent Southern Song artist Xia Gui (AD 1180 to 1230). (Photo. from Anon. (1953).)

pyramid, again presenting its narrowest axis to the current. The third method combines the two and is used when it is desired to lift the bridge well above the water level. Occasionally these supports were used for many spans.

Though unsuitable for single breadths exceeding 6 metres, stone beam bridges are to be found from time to time on the island waterways, including the Grand Canal, but in such cases the abutments extend far into the stream. The first bridge to be built in China wholly of stone, spanning a transport canal at Luoyang in AD 135, was very probably of this kind, though it may have had one or more arches. Among the many which remain in the cities of the eastern provinces, there are two certainly of the Song period at Shaoxing, one dating from AD 1256.

During the Song, there was an astonishing development: the construction of a series of giant beam bridges, especially in the Fujian province. Nothing like them is found in other parts of China, or anywhere outside China. These structures were (and are) very long, some of them more than 1 kilometre, and the spans are extraordinarily large, up to 21 metres, and involved the handling of masses of stone weighing some 200 tonnes. The art of building them was afterwards forgotten, and no record was kept either of

Fig. 444. Traditional rope suspended pile-driver, in use on the Old Silk Road. (Photo. Joseph Needham, 1943.)

the quarries which provided the stones or of the techniques by which they had been hauled to the site and set in position.

Much obscurity surrounds the name of the engineers responsible for these megalithic bridges, though some of them bear a record of the provincial officials under whose aegis they were constructed and repaired. But one may sense the existence of some master bridge-builder and a school or tradition founded by him. Certainly there is a famous folk-tale about this which cannot be omitted.

> Before the Pouyang [Luoyang] bridge was built, everyone had to cross over by a boat. In the reign of the Emperor Shen Zong (AD 998 to 1022) of the Song dynasty, a pregnant woman from Fuqing was crossing the river to Quanzhou; but just as the ferry was in the middle of the stream the bad tortoise and snake spirits sent suddenly a strong wind and high waves to upset and sink it. All at once a voice cried from the sky: 'Professor Cai is in the boat. Spirits must behave decently with him.' And scarcely had the word been spoken when the wind and waves died down. All the

Fig. 445. A stone beam bridge near Hangzhou. (Photo. from Mirams (1940).)

passengers had heard exactly what was said, but there was no one of that family in the boat except the pregnant woman from Fuqing. They therefore congratulated her, and she said: 'If I really give birth to a son who becomes a professor, I will charge him to build a bridge over the Luoyang river.'

Several months later she did bear a son, who was named Cai Xiang. Later on (about AD 1035) he did in fact become a professor. His mother related to him her experience on the river, and begged him to fulfil her vow. Being a dutiful son he immediately gave his consent, but at that time there was a law against anyone being appointed an official in his own province, and as Cai Xiang was a native of Fujian, he could not become governor of Quanzhou, which is in that province. Fortunately a friend of his, the Chief of the Eunuchs in the Imperial Palace, conceived a wonderful plan. One day, when it was announced that the Emperor would walk in the garden, he took some sugared water, and wrote on a banana leaf 'Cai Xiang must be appointed Governor of Quanzhou'. Ants immediately smelt the honey and gathered on the characters in vast numbers, to the stupefaction of the Emperor, who happened to pass the banana tree and saw them drawn up in the form of eight characters. The Chief of the Eunuchs watched him reading them over, and drew up a decree which the Emperor signed.

And so eventually the bridge did get built, with the help of various immortals.

But megalithic bridges, sea-walls and other public works in this province are connected particularly closely with the names of Buddhist monks, for whom bridge building was a beatific work and who were probably themselves engineers. The most famous of the group was Daoxun (*died* AD 1278), who built more than 200 bridges of various sizes in the province, and who is said to have completed what Cai Xiang had begun. His best work was near Quanzhou, but he also built dykes and sea-walls. So did Bofu in the following century (*died* AD 1330), and it is interesting to read that he always used to sleep, doubtless under very rough conditions, with the workers 'on the job'.

Some idea of these bridges may be gained from Figs. 446 and 447. These illustrate the Jiangdong Bridge, on the post road from Guangdong to Quanzhou, and the Wan'an, or Luoyang, Bridge north-east of that city, details of which, together with some of the other bridges of the region, are given in the accompanying table (Table 55). Both Marco Polo and John of Monte Corvino must have crossed these bridges. The number of spans in

(a)   (b)

Fig. 446. A megalithic beam bridge in south-east China; the Jiangdong Bridge across the Jiulong River, upstream from Amoy in Fujian. (a) Stone beam span at the sixth pier, from the north-east; (b) another pier from upstream (north-west). (Photo. from Ecke (1929).)

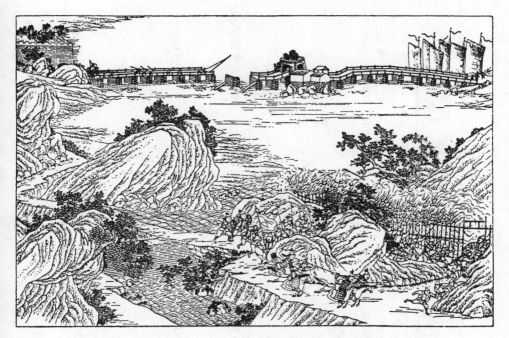

Fig. 447. A megalithic beam bridge; the Wan'an bridge across the Lu
River, north-east of Quanzhou in Fujian. Built between AD 1053 and 1059,
its length is 1097 metres. A Chinese drawing from the *Wu Jiangjun Tu
Shuo* [*Wu Chiang-chün Thu Shuo*] (Illustrated Account of the Exploits of
General Wu) (*c.* AD 1690) by Wu Ying. Wu was a military and naval
commander on the Qing side in the fight against the Ming forces in the
1660s and 1670s. In these campaigns the bridges had much strategic
importance. The mêlée depicted in the foreground is an episode from Wu's
exploits in 1678, while in the background repairs to the bridge are
proceeding.

relation to the total length of these Fujianese structures is very variable
because, as the huge stones of the deck failed, later generations were unable
to replace them, and had to insert new piers between the old ones. In many
cases the piers are corbelled or cantilevered out from their bases to give
additional support to the trusses, and the piers themselves are in general
boat-shaped. The largest beams exceeding 200 tonnes in weight and those
which Periera so much admired (p. 116) have cross-sections 1.5 metres high
and 1.8 metres broad.

With bridges of this type, the greatest difficulty was with the foundations,
which in many cases proved insufficient, so that in different places in Fujian
a number of ruined giant beam structures are to be found. It is clear that,

Table 55. *Some megalithic beam bridges of Fujian.*

Bridge (Qiao, abbreviated to Q.)	Location	River (jiang) spanned	Date of construction (all dates are AD)	Total length (metres)	Breadth (metres)	No. of spans	Greatest span length (metres)	Weight of largest beams (tonnes)	Builder
Wanshou Q.[1] or Wenchang Q.[5,a]	Fuzhou[2] (Minhou[6])	Min jiang[3]	1297 to 1322	625 (including shorter continuation south of Nantai[7] Island)	4.4	36	14	c. 81	Wang Fazhu[4]
Hongshan Q.[8]	a little further up river	Min jiang[3]	1476	799	4.4	46	–	–	–
				–	–	28	–	–	–
Fuqing Q.[9]	Fuqing	local tidal estuary	between late Song and early Ming	c. 244	–	17	c. 8	–	–
Luoyang Q.[10] or Wanan Q.[12]	north-east of Chuanzhou[11]	Luoyang jiang	1023, or, more probably, 1053 to 1059	1097	4.6	47 (now 121)	c. 20	c. 152	Cai Xiang
Panguang Q.[13] or Wuli Q.[14]	north-east of Chuanzhou[11]	Luoyang jiang	c. 1255	>1220	4.9	–	–	–	Daoxun
Shunzhi Q.[15]	south-west of Chuanzhou[11]	Jin jiang[16]	1190 to 1211 (repaired 1341, 1472 and c. 1650)	457	–	–	–	–	–
Fou Q.[17]	south-west of Chuanzhou[11]	Jin jiang[16]	c. 1050 as a floating bridge; in stone, 1160	244 (now strengthened to take a main road)	5	130	c. 8	–	Fachao

Anping Q.[18] or Wuli Q.[14]	Anhai[19]	local tidal estuary	12th century	1524	360	4.6	–	–	–
Tongan Q.[20]	Tongan	local tidal estuary	1094, repaired 1294	305	18	18	–	–	–
Jiangdong Q.[21] or Qiandu Q.[24] (='Bolam Bridge')	32-kilometres up river from Amoy, east of Zhangzhou[25]	Jiulong jiang[22]	1190 as a floating bridge; stone piers 1214 and beams 1237	335	19	5.5	>21	>203	Li Shao[23]
Tongjin Q.[26] or Laoqiao Q.[28]	south of Zhangzhou[25]	Long jiang[27]	end 12th century	274	28	12	–	–	–
Guangji Q.[29] or Xiangzi Q.[33] (in Guangdong)	Chaozhou[30]	Han jiang[31]	Song, 1170 to 1192	497	21	6	–	–	Ding Jiuyuan,[32] Shen Zongyu,[34] et al.

(This has an 82 metre pontoon segment to permit the passage of large ships)

[a] As in private duty bound, Dr Needham celebrates the Yuan builder of this most impressive bridge, the only one of the group that he himself has seen, remembering that sunny May morning of 1944 when Dr Huang Xingzong and he were borne in rickshaws over its granite baulks to spend the first of several enjoyable days among the bookshops and hot springs of the city of Fuzhou. The greater part of his collection of the mathematical classics was purchased there.

1 萬壽橋   2 福州   3 閩江   4 王法助   5 文昌橋   6 閩侯   7 南台   8 洪山橋   9 福清   10 洛陽橋
11 永州   12 萬安橋   13 盤光橋   14 五里橋   15 順治橋   16 晉江   17 浮橋   18 安平橋   19 安海   20 同安橋
21 江東橋   22 九龍江   23 李韶   24 虔渡橋   25 漳州   26 通津橋   27 龍江   28 老橋頭   29 廣濟橋   30 潮州
31 韓江   32 丁久元   33 湘子橋   34 沈宗禹

whoever was mainly responsible for them, the chief period of their construction was the eleventh to thirteenth centuries AD.

In modern times, studies have been made on the strengths of stone bars similar to those used in medieval China. Using sample lengths varying from 1.5 to 3 metres and from 15 to 46 centimetres in cross-section, the tensile or breaking strength was found; grey granite proved to be almost 2.3 times stronger than red granite. Taking the weight of the stone and a likely maximum for the highest superimposed load on the deck of the bridge, it is possible to calculate the upper limit of length for a single beam; using the best figures, the result was 22.5 metres, exactly corresponding to the maximum lengths of the Fujian bridges. The Song builders therefore reached the utmost limit of practicability, for baulks of any greater length would break under their own weight. How they reached their result we do not know; possibly by bitter experience of failures, or perhaps by actual tests of the strength of the stone at the quarries.

The Fujianese type of bridge building gives an impression of uncouth strength and determination, with a prodigious waste of material, as regardless of cost as the most massive Roman masonry. Yet it was a purely provincial style, for, as we shall see, six or seven centuries earlier the builders of arch bridges in Northern China had already attained an economy of materials and a grace of design which was not approached in Europe until the dawn of the Renaissance. Nevertheless, the giant beam-bridge is not to be despised, and holds its own for certain purposes to the present day.

Before leaving the realm of the beam, a few moments must be given to drawbridges and pontoons. Drawbridges did not figure prominently in fortifications during the Qing (AD 1644 to 1911), bridges over moats just being made easily dismountable at need. But in earlier times mechanically movable bridges had been used a good deal. On the journey of the Spanish envoys from Amoy to Fuzhou in AD 1575, Miguel de Loarca described a great bridge of many spans outside Quanzhou, which had a drawbridge at its end. But nearly 1000 years earlier, the great Daoist military encyclopaedia of AD 759, the *Tai Bai Yin Jing* (Manual of the White (and Gloomy) Planet (of War; Venus)) had an entry for the subject in its section on the defence of cities. Its author Li Quan wrote:

> Turning Portal Bridge. For this kind of bridge flat planks are used as the deck. At (one) end of the bridge there is a horizontal bolt, and when this is removed the (whole) bridge turns away from the gate, so that soldiers and horses cannot pass over, and all of them fall into the moat. (Formerly the King of) Qin used such a bridge as this (in the attempt) to kill Dan, Prince of Yan.

The story about Prince Dan occurs in the ancient fictional biography *Yan Dan Zi* ((Life of) Prince Dan of *Yan*), written about the end of the second century AD. The prince, escaping in 232 BC from the ruler of Qin State, where he had been ill-treated as a hostage, had to cross a bridge equipped with some release mechanism whereby the king intended to kill him, but he passed over before it could operate. Unfortunately the mechanism of these ancient 'drawbridges' is not clear from Li Quan's description. They were evidently not raised in the air like the drawbridges of medieval Europe, but it seems clear that they either turned over sideways (which might have been rather effective), or else swung on hinges so as to drop into the moat probably at the end nearer the gate and wall. Later on (p. 153), we shall refer to a submersible suspension bridge of the fifth century AD which acted as a drawbridge.

Floating pontoon bridges are of high antiquity in Europe since the Greek Mandrocles of Samos is credibly reported to have constructed one over the Bosphorus for an expedition of Darius I against the Scyths in 514 BC. But this date may well be anticipated by the first Chinese reference, which takes us back to a text of the eighth century BC. Let us listen to Gao Cheng discussing it in the *Shi Wu Ji Yuan* (Records of the Origins of Affairs and Things) of about AD 1085.

> The *Chun Qiu Hou Zhuan* (A History of the Ages since the Time of the *Spring-and-Autumn Annals*) says that in the fifty-eighth year of the Zhou High King Nan (257 BC), there was invented in the Qin State the floating bridge with which to cross rivers. But the Da Ming ode in the *Book of Odes* says (of King Wen) that he 'joined boats and made of them a bridge' over the River Wei. Sun Yan comments that this shows that the boats were arranged in a row, like the beams (of a house), with boards laid (transversely) across them, which is just the same as the pontoon bridge of today. Du Yu also thought this. . . . Zheng Kangcheng says that the Zhou people invented it and used it whenever they had occasion to do so, but the Qin people, to whom they handed it down, were the first to fasten it securely together (for permanent use).

For the Qin State's bridge of boats across the Yellow River there is much better authority. Sima Qian himself recorded its building by Prince Zhao Xiang of Qin at the date given above. With periodical renewals, this Pujin Bridge, as it came to be called from its geographical position near the border of Shanxi north of the great bend at Tongguan, lasted for very many centuries, the senior and most famous of the three great floating bridges across the Yellow River. The *Chu Xue Ji* (Entry into Learning) encyclopaedia of

AD 700 mentions it, and we have already found the Japanese monk Ennin admiring it in AD 840. He estimated its length at about 305 metres.

From the eighth century AD also we have much information about the upkeep of these important bridges, notably in the manuscript fragment of the Tang government ordinances of the Department of Waterways (AD 737). Here we learn of the bridgekeepers, watermen and maintenance artisans kept permanently on duty and exempt from military or other service, active in the defence against dangerous floating lumber in flood time and watchful to undo all fastenings when the ice set hard. Replacement pontoons were built in special shipyards, and up to half the total number were kept available, while provision was made for the manufacture, storage and periodical testing of the necessary bamboo hawsers. The Pujin Bridge was always an important strategic crossing.

The mention in the quotation above of Du Yu, the eminent engineer of the Jin period (AD 222 to 284), is interesting because he himself constructed the second famous pontoon across the Yellow River, the Heyang Bridge north-east of Luoyang. It must have been quite a feat, for many of the imperial advisors urged that the plan was impossible.

References to pontoon bridges in the Tang and other periods are not infrequent. For example, during the rule of the Jin Tartar dynasty, Zhang Zhongyan built one in AD 1158, while Tang Zhongyu used iron chains for securing one in 1180; another, made using inflated skin rafts, was constructed by Shimo Anzhe during the next dynasty. Many were mentioned in the account of the travels of the Daoist Qiu Changchun in about 1221, and a famous one across the Amy Darya river was constructed in a single month for Genghis Khan's second son. In the fifteenth century AD, there was another built at Lanzhou, which greatly impressed foreign travellers.

Dr Needham found that a special type of pontoon bridge occurs on the western border of Sichuan, near Yazhou, where at certain times of year a bamboo raft is moored afloat from shore to shore, and across it a gangway, at least for pedestrians, is laid. Bridge pontoons are also shown in Ming technical books.

All this, however, has thrown no light on the oldest use of the device. Though the *Book of Odes* certainly refers to the founder of the Zhou dynasty, Wen Wang in the eleventh century BC (p. 91), to insist dating it at that time would be most unwise; the eighth to seventh centuries BC would be quite enough to do it justice. Its phrase about the bridge echoes down through the Han period, as in the *Dong Du Fu* (Ode on the Eastern Capital), written by Ban Gu in AD 87, and much later still. There can thus be no doubt that the pontoon bridge is a very old institution in China, and since the ode in the *Book of Odes* must date from the first half of the first millennium BC, it looks as the 'artisans of King Wen', whoever they were, took precedence over

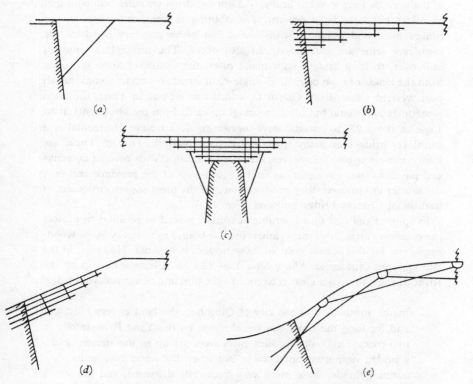

Fig. 448. Types of cantilever bridges. (*a*) Simplest type of strainer-beam
support, used in small bridges. (*b*) Horizontal cantilever, used over river
gorges. (*c*) Combination of horizontal cantilever and strainer-beam supports,
often found in the many-spanned bridges in Hunan. (*d*) Soaring cantilever,
used across gorges or routes of water-traffic. This form occurs also in
Tibet. (*e*) Multi-angular soaring cantilever, in which the main beams spring
forth at two or more angles with the abutments. (See also Fig. 450.)

Mandrocles of Samos. But it would not be at all surprising to find that
Babylonian skill outdated them both.

Cantilever bridges

In all southern and western parts of China, and especially near the Himalayas and Tibet, the people responded to the challenge of gorges and tur-
bulent streams up to 46 metres wide by the construction of cantilever bridges.
The two cantilever arms were built out with superimposed timbers from the
sides in various ways (see Fig. 448), and were then connected in the middle

of the span by long wooden beams. Additional struts (strainer beams) might or might not have been present. The abutments were, and often are, of timber filled and weighted with stones, but heavy masonry in which the cantilever arms are embedded can also occur. The principle is applied, moreover, to long bridges with many piers, the cantilever arms springing from the heads of each pier. Bold single-span structures occur commonly all over western China from Gansu to Yunnan as well as in Tibet, the arms sometimes horizontal but often 'soaring' upwards from the abutments at an angle of about 25° to join the level connecting deck beams. Horizontal-arm cantilever bridges of many spans occur particularly in Hunan. These are impressive enough in themselves, but when combined with covered housings and pavilions over each pier, as in the south-west of the province and over the border in Guangxi, they produce some of the most superb structures of traditional Chinese bridge building (Fig. 449).

In spite of the fact that boarding is usually placed in position to protect the cantilever arms from the weather, there is bound to be decay in the wood-work, and for this reason none of these bridges is very old. However, in the late eleventh-century AD *Sheng Shui Yan Tan Lu* (Fleeting Gossip by the River Sheng), there is a clear reference to the building of a cantilever bridge:

> In the south-west of the city of Qingzhou the land is very hilly, and for long the town was cut through by the Yang River into two parts. Originally wooden piers were set up in the stream, and a bridge supported upon them, but when the water rose in its autumn floods, these piers were frequently damaged, and the bridge became unsafe. The magistrates were always worried about this. In the Mingdao reign-period (AD 1032 to 1033) the governor of Qingzhou, Xia Yinggong, was greatly desirous of overcoming the difficulty (and gave encouragement to) a certain retired prison guard, who was known for his ingenuity. He piled up large stones to make firm abutments, and then by connecting several dozen great beams together, he threw across the river a kind of 'flying bridge' with no central pier. Though now more than fifty years in use, this bridge has never suffered injury. Then in the Qingli reign-period (AD 1041 to 1048), when Chen Xiliang was governor of Suzhou, he noted that the bridges on the Bian Canal often fell into disrepair, so that they damaged or destroyed official shipping, hurting and even killing travellers. He therefore ordered that (they should be rebuilt) following the pattern of the Flying Bridge of Qingzhou. Now all the bridges between the Feng River and the Bian Canal are of this type, which is a great benefit to communications. The common people call them 'rainbow bridges'.

Fig. 449. Horizontal cantilever bridge of four spans over the Linqi River north of Shanjiang in northern Guangxi. Its structure is notable for the elaboration of features used elsewhere, the crowning of piers by pavilions with many-tiered roofs, and the provision of a spacious roofed gallery over the deck. (Photo. from Anon. (1849).)

The cantilever principle was thus most welcome, since it obviated the necessity for central piers, obstacles particularly liable to flood damage and prone to get in the way of navigation. Forming veritable illustration of this passage is the depiction of a great bridge outside Kaifeng, the capital of the Northern Song by Zhang Zeduan about AD 1125 on the very eve of its capture by the Jin Tartars. With its wonderful detail of everyday life, this *Going up the River at the Spring Festival* has already helped us (in chapter 3, p. 83), and here Fig. 450 shows a many-angled soaring cantilever type bridge. It is borne not only upon about ten great beams arising out of the abutments at some 40° and supporting a series of horizontal members, but also upon another set interspersed within them and rising at some 55° to sustain corresponding pairs of sloping members which meet at the crown of the structure. The whole is trussed together with bars and collars similar to those used with the more ordinary bundles of parallel cantilever beams.

The further historical explanation of these beautiful forms of bridge building is not a very easy matter. A clear description of a timber cantilever bridge with stone abutments is given in the *Shazhou Ji* (Records of Shazhou) of the fourth or fifth centuries AD; it refers to one then built by the Tuyuhun people (a Xianbi tribe) across the river of the Dunhuang oasis. This seems to be the oldest textual record. Tradition speaks of a cantilever bridge in the Tang (AD 618 to 907), but by the time of the Ming (sixteenth century AD) the technique which produced the Kaifeng bridge seems to have been lost (or was at any rate unknown to non-provincial scholars), for Ming copies of Zhang Zeduan's painting replace the cantilever structure by a single great arch. But possibly in the meantime its simple lattice-like geometry had travelled to Europe, for an almost identical design was proposed by Leonardo da Vinci (*c*. AD 1480) for a military bridge. Searching back in Chinese literature, much will depend on the extent to which the term 'rainbow bridge,' mentioned at the end of the previous quotation, is to be taken as a technical rather than a popular descriptive term. For example, in the late third century AD Zhou Chu wrote in his *Feng Tu Ji* (Record of People and Places): 'In front of Yangxian there is a great bridge 219 metres long from north to south, and very high in the middle so that it looks like a rainbow.' Was this a series of short spans of parallel cantilever pattern with a single

Fig. 450 The great bridge at Kaifeng from the scroll painting *Going up the River at the Spring Festival* made about AD 1125 by Zhang Zeduan. (Photo. Zheng Zhenduo [Chêng Chen-To].) So far as is known, no examples of this multi-angular soaring cantilever construction still exist in China, but in pre-Ming times there seem to have been many. A towpath gallery can be seen under the bridge along the further abutment, and the stern sweep of a great barge which has just passed under the bridge appears on the near side of the river.

Fig. 451. Han stone relief of a pavilion built out over a lake and supported by corbel bracket-arms on the cantilever principle. Preserved at the Temple of Confucius at Qufou in Shandong. (Photo. Joseph Needham, 1958.)

soaring, perhaps multi-angular, cantilever span in the middle to allow the passage of traffic under sail, or a series of arches with a central arch?

One fact possibly relevant to the origin of the cantilever bridge in China is that in Han times pavilions were sometimes built out over lakes on corbelled brackets resembling cantilevers. Such structures were called *ti qiao* [*thi chhiao*] (ladder bridges), and in 1958 Dr Needham photographed a stone relief of one in the Temple of Confucius at Qufou in Shandong (Fig. 451). In 1971, a pavilion somewhat of this kind still existed at the Yan-Yu Temple at Hanoi in Vietnam, dating from AD 1049. It is carried on a cylindrical stone column rising out of a pool and upheld by wooden cantilever corbels and beams. But the roof-supporting system of corbel brackets in Chinese architecture as a whole, and not only such corbels as these, presents itself as the background of the Chinese cantilever bridge.

Such bridges extend outside China, and a bridge thrown by Trajan across the Danube in AD 104, which is depicted on his famous column of AD 113, shows what seems to have been a cantilever bridge of many spans on stone piers; with a computed distance of 52 metres for each span that may not have been impossible. Yet if so, it seems to have been an isolated effort in western antiquity. It would therefore be interesting to know the basic home of the original design, and the stages of its spread, but the invention seems to have taken place so early that it is difficult now to trace its course of history. Perhaps if the Himalayan region was the zone of origin of both cantilever and suspension bridges, we might expect them to occur combined, and this does occasionally happen. It is also possible, though in China unusual, to build a cantilever bridge in stone – there is an example in Zhejiang province. This brings us to the arch.

Arch bridges

The generally accepted view used to be that the arch was an Etruscan invention of perhaps the fifth century BC, and that its Italian origin accounted for the fact that the Romans made such great use of it while the Greeks employed it hardly at all. Argument has also centred round the use of the arch in India. All such discussions have now entered a different phase with the establishment of the fact that the arch, the vault and the dome were also familiar to Sumerian Mesopotamia. This makes it easier to understand the appearance of the arch and barrel vault in full flower in Qin and Han tombs, and to suppose that the use of the arch for city-gates and bridges was well understood by the Zhou people, if not indeed by the Shang.

In antiquity, both in West and East, arches were invariably semicircular. This type of arch will not conveniently bridge gaps of more than about 46 metres, and the average spans of Roman semicircular bridges range between 18 and 24 metres, while aqueduct arches are generally much less, about 6 metres. The longest surviving Roman arch spans almost 36 metres.

The essentials of the arch itself, with its ring of shaped tapering stones culminating in the keystone at the crown, were naturally identical in Roman and Chinese work. Yet the bridge engineer H. Fugl-Meyer found, when he examined Chinese methods, that the whole construction was so unlike in the two kinds of work that he was convinced that there had never been any contact between Roman and Chinese bridge-builders. The Roman arch bridge is of very massive masonry, so bulky that in some cases the chalk in the middle of the structure has remained plastic to the present day and will still harden when exposed to the air. It was also over-dimensioned, and consequently unnecessarily wasteful of material. On the other hand, the characteristic Chinese bridge arch is a thin stone shell loaded with loose filling, and this is kept in place by side-walls of stone, and topped with stone slabs to form the deck and approaches. While such a method was most economical in materials, it was also much more liable to deformation caused by any rise or fall in the foundations, and the Chinese were therefore driven to invent

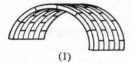 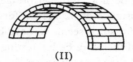 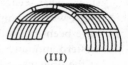

(I) (II) (III)

Fig. 452. Types of stone bonding in bridges. (After Mao Yisheng and Luo Ying.)

a number of subsidiary devices, bonding stones to hold the casing together, and built-in 'shear walls' running through the structure to act against deforming forces. Fugl-Meyer commented 'Since the Chinese bridge was constructed of a minimum of material it was an ideal engineering product, fulfilling both technical and engineering requirements.'

For the arches themselves, several different types of bonding were used as the stones were laid down on the wooden centering (Fig. 452). In the transverse method (*bing lie* [*ping lieh*], 'abreast') a number of essentially separate arches were built up side by side in a series till the required width of the bridge was reached (Fig. 452(I)). The stones were often elongated like arcs of a circle, and the adjacent arches were always bonded by alternate jointing. This was the method used for the great 'segmental arch' bridges to be mentioned shortly; it had the advantages that the collapse of one or two particular arches would not put the bridge out of action, and that as the building of each arch required only a narrow centering, this could be moved on as each arch was completed. Thus there was a great saving in timber. The longitudinal method (*zong lian* [*tsung lien*], 'lengthwise linked') was different (Fig. 452(II)); the shaped stones were built up row by row from both end supports of the arch, alternating from side to side like headers and stretchers in brickwork, over the whole width of the bridge. A third method (Fig. 452(III)) was in a way a combination of these two (*lian suo* [*lien so*], 'chain-linked'); here, long curved arc stones were prepared according to the radius intended, and made to alternate with tie stones forming part of the arch lining but extending fully across its width. This was more suitable for small bridges. The arch rings of old Chinese bridges can bear a greater load than modern stress analysis of the arch alone would seem to justify, because the walls and fillings of spandrels (the spaces between the arches) and abutments exert a 'passive pressure' on the arch itself, increasing its strength and rigidity. Moreover, the wooden centering was traditionally given a little 'camber' so that when it was removed the stones sank under their own weight and pressed the arch more closely together. Three other combinations of types (I), (II) and (III) were also adopted.

Mortar was seldom used in the arch or side walls, and never in the foundations, as the composition of hydraulic mortar was not known. The stones were, and are, mortised into each other with crossets (where a projection fits

Fig. 453. An arch bridge at Kunshan in Jiangsu, showing the ends of the
shear walls in the piers. (Photo. from Mirams (1940).)

into a recess), or held together with iron cramps in double dovetail shape,
sunk into indentations of the same shape cut in the stones.

In the side walls between the arches, square ends of stone can often be
seen; these are the 'through-binders', or long stone ties running from wall
to wall to prevent the walls themselves from bulging outwards. Since in
many parts of China, such as the Yangtze delta, the subsoil was so bad that
it was impossible for Chinese engineers to make foundations unyieldingly
firm, the built-in shear wall was devised. This is a buried vertical stone wall
running through the filling each side of the arch at right-angles to the bridge
itself, as can be seen well in Fig. 453. Such a wall is made of long stone slabs
placed vertically side by side, mortised into the foundation and gathered at
the top into a long horizontal stone which ties once again with the stone
walls. In essence, they are identical to the simple rows of vertical slabs which
constitute one of the forms of pier for the smaller kinds of stone beam
bridges (p. 119); but their combination with the arch was a brilliant idea.
For while the arch itself is a loose chain which cannot resist any bending
force, those forces tending to deform the arch are transferred through the
clay and chip filling to the two shear walls. Indeed, an arch can be deformed
until it is almost elliptical and still do its duty.

Another remarkable invention of Chinese bridge engineers was that of the complete circle structure, the arch of the bridge above being mirrored by a corresponding inverted arch springing from the abutments deep under the water. Great stability, especially valuable when rock or clay foundations are not to be had, can be obtained by such rings of masonry. The first bridge of this kind in China was built between AD 1465 and 1487 near Wuxian in Jiangsu province, and was still in active use in 1971.

Travellers in China have always remarked on the frequency of the very humpbacked decks and approaches, but this was a natural feature in a civilisation which depended a good deal on porters and pack animals, and appears noticeably also in medieval European bridge building. One obvious reason was to facilitate the passage of ships without lowering their masts and sails; a point specifically mentioned by Marco Polo.

Very few arched bridges, if any, remain from the Han and San Guo (Three Kingdoms) periods. However, the *Sui Jing Zhu* (Commentary on the Waterways Classic) of the late fifth or early sixth century AD has preserved an account of a great stone bridge at Luoyang which distinctly mentions a lofty arch; this was built towards the end of the third century AD. The author, Li Daoyuan, says:

> As for the many bridges (of Luoyang) they are all made of (dressed) stones piled up into high and gallant structures. Although with the passage of time there is some decay, they never fail to do their office. When Zhu Chaoshi was travelling (in his military campaigns) he wrote to his elder brother saying, 'Outside the palaces of Luoyang some six or seven *li* away there is a bridge built all of great blocks of stone, and underneath it is rounded so that not only does the water pass, but also large ships can go through. An inscription on it says that it was built in the third year of the Taikang reign-period (AD 282). In the construction over 75 000 men were employed each day, and after five months it was finished.' In the course of years this bridge fell into disrepair and was then completely restored, but the inscription is no longer to be seen.

Such was the Luren Bridge or 'Bridge of Wayfaring Men'. Chinese historians of engineering have found no textual evidence for arched bridges earlier than this, but if the one here described had a span of some 24 metres like the Pons Fabricius, as would seem to follow from what is said of the river traffic, it cannot possibly have been the first of its kind, and we could reasonably conclude that small arched bridges must have been constructed first in the Later Han period, if not before.

The Tianjin bridge at Luoyang was reconstructed with streamlined piers towards the end of the seventh century AD by Li Zhaode, a renowned engineer who built many other bridges at the same capital. This century is indeed something of a turning point, for while hardly any existing structures can reliably be considered earlier in date, some of the noblest and most interesting that China possesses undoubtedly derive from its initial decades. We can therefore entertain no doubt that between Ma Xian (*flourished* AD 135) and Li Chun (*flourished* AD 610) there was a powerful and vigorous tradition of arched bridge building, though few or even no identifiable examples have happened to survive.

The major works of arched bridge building in China are, of course, those in which many spans had to be constructed. Beautiful bridges of three or four arches exist in every city, but pre-eminent among those with long arcades is the Wannian Bridge at Nancheng in Jiangxi, partly because a special book was devoted to it, the *Wannian Qiao Zhi* (Record of the Bridge of Ten Thousand Years) written in AD 1896 (see Fig. 454). Crossing the Rushui River, the bridge had twenty-four arches and was 550 metres in length. From literary sources we know that previously, between AD 1271 and 1633, crossing had been made by a pontoon bridge, but the drowning of some thirty people in this latter year led to the building of the multi-arched bridge, which was completed in 1647. From then on it was always an important link between Jianxi and Fujian.

Here we may conveniently refer to the use of metal dowels and cramps in Chinese stonework. 'Mutual inlaying' was common practice from the Tang (AD 618 to 906) and Song (AD 960 to 1279) onwards. Certainly, iron cramps were used in an arch bridge of particular interest erected early in the seventh century AD, as we shall see in a moment (Fig. 454(*d*)), but before that point it is difficult to trace back. One would hardly expect to find it in the stone tomb-shrines of the Han, which were not subject to any great stress, and no stone bridges of that period have survived to provide information.

The greatest invention in arched bridge building, both for engineering merit in economy of materials and for aesthetic quality, came when engineers dared to abandon the idea that security demanded semicircular arches in which the line of the pier approached the curve like a tangent and so 'conducted the weight vertically downwards'. When it was realised that the curve of the arch could be made much flatter provided it was a segment of a much larger semicircle, and that the bridge could thus be made to fly forth from its abutments as if tending to the horizontal, the segmental bridge was born. This discovery, which in Europe may have had something to do with the parallel development of the flying buttress, seems to have been made by Westerners late in the thirteenth century AD, but it was not applied widely and daringly until the fourteenth century. The height above the chord-line (joining the lower ends of the arch) was gradually reduced from the full radius of a circle to less than one-half (see Table 56, pp. 148–9).

The notable Western bridges of this kind are rightly regarded as great achievements, yet it must remain an extraordinary fact that a comparable

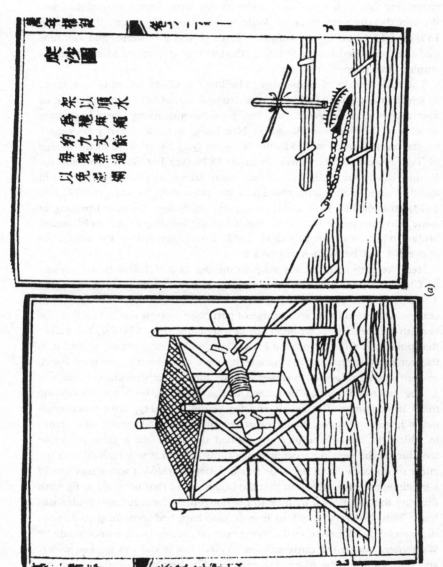

Fig. 454(a). For caption, see p. 145.

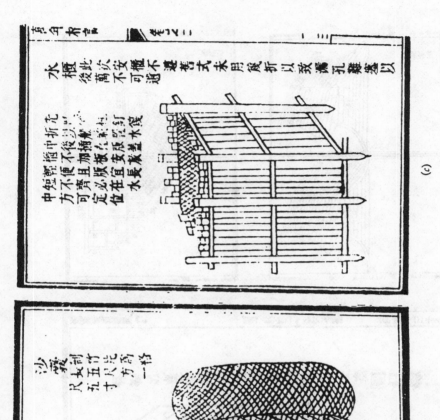

(c)

水
櫃
後
此
兩
次
安
櫃
不
遷
者
式
未
用
度
折
以
效
滿
孔
難
塞
以

中短密爾巾折亡
方不便不後多
可謂且加捅松
定必版板名彩出
位在版宜支版浮釘
水長夜基复基水度

(b)

沙
霙
尺大訓
五只只西
寸方先方
一格

Fig. 454(b),(c). For caption, see p. 145.

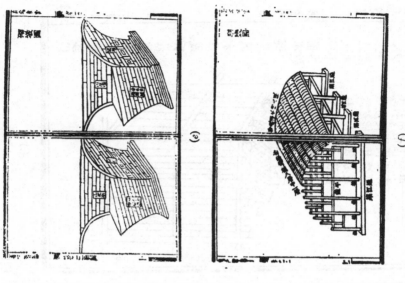

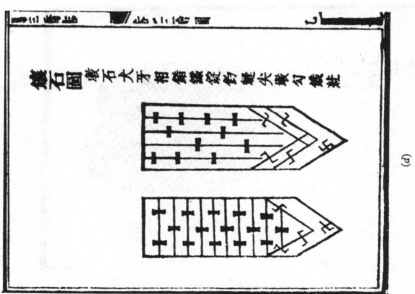

Fig. 454(d),(e),(f).

bridge of even more advanced character, together with a number of smaller ones, were built in about AD 610 by a Chinese engineer of outstanding quality, Li Chun. His activity and that of his pupils – for it is clear that he founded a school and style which lasted for centuries – left its mark mostly in the provinces of Hubei and Shanxi, centring on his finest work, the Great Stone Bridge named Anji near Zhaoxian. The terrain here is at the edge of the North China Plain at the foot of the Shanxi mountains, and the river across which Li Chun threw his bridge is the Jiao. The structure is seen in Fig. 455, and Chinese engineers justly regard it as one of the greatest achievements of their ancestors. A single arch bridge with a span of 37.4 metres, rising 7.1 metres above the chord-line, it even surpasses the galaxy of European bridges of the fourteenth century, for its spandrels are perforated by two arches on each side. Its design and construction not only reduced the resistance to the flow of water in floods, but also lessened the burden on the arch, as well as economising in materials.

Li Chun's bridge still carries inscriptions written by Tang officials and later visitors praising the 'system of four holes near two banks'. One, from about AD 675, says:

This stone bridge over the Jiao River is the result of the work of the Sui engineer Li Chun. Its construction is indeed unusual, and

---

Fig. 454. Illustrations from the *Record of the Bridge of Ten Thousand Years of* AD 1887 by Xie Gantang. (*a*) The dredger (*pa sha*). 'The winch mounting should be set up in conformity with the current. The hemp cable more than 90 ft. long, should be heat-dried every night to avoid damp rot.' (*b*) The gabions (*sha nang*). 'Bamboo strips are woven into loose nets 5 ft. long and 1.5 ft. across.' (*c*) The cofferdam (*shui gui*). 'This is a divergence from old methods; we did not use bamboo matting (alone), because it is extremely difficult to stop leakage in that way – that is a point not to be forgotten. First the main framework is set up, and then the bamboo matting nailed on; in this state it is light and convenient to handle, and a boat takes it out to its position. Later (after draining) wooden boards of unequal length are inserted within the framework; it is only when it is in position that the lengths of the boards can be fixed (to fit the inequalities of the river-bed).' (*d*) Dowelling and cramping the stones of the piers (*xiang shi*). 'The pier stones are inlaid with metal alternately, like dog teeth, and iron anchor pieces (dowels) hook the seams together. At the pointed (upstream) ends, the stones are fastened with hooks (cramps).' (*e*) The piers (*qi dun*). See text. (*f*) The half-centring (*pian weng*). 'Each side has eight "rafters". The curvature changes in eight steps.' Notable here is the use of pillars, cross-beams, tie-beams and king- and queen-posts, exactly as in the transverse framework of a Chinese building, but with the cross-section of the array of purlins arranged to be convex rather than concave. (The two last illustrations are full double-page openings like (*a*), but have been reduced to half-scale.)

Fig. 455. The oldest segmental arch bridge in the world; the Anji bridge across the Jiao River near Zhaoxian in southern Hubei, built by Li Chun about AD 610. (Photo. from Mao I-Shêng (1961).)

no one knows on what principle he made it. But let us observe his marvellous use of stone-work. Its convexity is so smooth, and the wedge-shaped stones fit together so perfectly. . . . How lofty is the flying-arch! How large is the opening, yet without piers! . . . Precise indeed are the cross-bondings and joints between the stones, masonry blocks delicately interlocking like mill wheels, or like the walls of wells; a hundred forms (organised into) one. And besides the mortar in the crevices there are slender-waisted iron cramps to bind the stones together. The four small arches inserted, on either side two, break the anger of the roaring floods, and protect the bridge mightily. Such a master-work could never have been achieved if this man had not applied his genius to the building of a work which would last for centuries to come. . . .

In the Ming, Wang Zhijian, the author of the *Biao Yi Lu* (Notices of Strange Things) said that the bridge looked like a new moon rising above the clouds, or a long rainbow hanging on a mountain waterfall. Ambassadors and other important travellers used to go out of their way in order to see it. Indeed they still do. After the time of Li Chun, spandrelled arches were used in many twelfth century AD Chinese bridges, but they did not appear in Europe until the fourteenth century. Li Chun's spandrel arch construction was thus the ancestor of those bridges of reinforced concrete which dispense with all filling between the arch and the deck, connected only by vertical pillars or a network of concrete members.

In order to appreciate the full originality of Li Chun, we must take a brief look at the statement just made that the invention of segmental arches appeared in Europe no earlier than the fourteenth century AD. This is true only if we refer to free-standing bridges, for segmental arches embodied in buildings go back to the Hellenistic period. They are present in the temple of Deir al-Medineh in Egypt, built in Ptolemaic times (early second century BC). Surprisingly, it may turn out to be characteristic of ancient Chinese brickwork too, for traces of two built-in segmental arches or vaults have come to light in the tomb of Liu Yan, the Prince of Zhongshan who died between 88 and 90 AD, as well as in the larger vaulted tomb of similar or earlier date at Yingchengzi. But what could be done in building was not so easy to apply in bridges, which had to launch forth into the void.

The attempt has sometimes been made to show that many medieval bridges were segmental and that the departure of the fourteenth century AD was not so revolutionary. Fortunately, this question is susceptible to measurement. A simple measurement of the 'flatness' of an arch can be obtained by calculating the ratio of the rise to the half-way point along the span, $s/\frac{1}{2}l$, where s is the sagitta (which in the case of a semicircular arch is the radius)

Table 56. *Segmental arch bridges in East and West.*

Date of construction	Province	Place	Bridge (Qiao abbreviated to Q.)	l (metres)	s (metres)	s/½l	l²/s
			semicircular arches	6	3	1.0	12
			semicircular arches	15	7.5	1.0	30
			semicircular arches	30	15	1.0	60
62 BC	—	Rome	Pons Fabricius	24.8	10.3	0.83	60
AD 1187	—	Avignon	Pont d'Avignon	33.7	14.0	0.83	81
14th century AD (main part, AD 1245)	—	Lyon	Pont de la Guillotière	33.2	11.6	0.70	95
AD 1285 (finished AD 1305)	—	nr. Bollène	Pont St Esprit	34	8.7	0.51	132
AD 1345	—	Florence	Ponte Vecchio	29.9	5.6	0.37	160
AD 1351	—	Pavia	Ponte Coperto	c. 30	6.1	0.41	147
AD 1354	—	Verona	Castelvecchio	48.5	13.7	0.57	172
AD 1375 (destroyed AD 1417)	—	Trezzo	Visconti Bridge	74.0	20.7	0.56	264
AD 1404	—	Sisteron	Pont de Castellane	28.0	7.1	0.51	110.4
c. AD 1525	—	Florence	Michael Angelo's single-span bridge	42.4	8.6	0.41	209
AD 1569	—	Florence	Santa Trinità	32.3	5.8	0.36	180
AD 1591	—	Venice	Rialto	6.9	6.9	0.48	118
AD 610	Hubei	Zhaoxian[1]	Anji Q.[2]	37.5	7.2	0.38	195

Province	County	Bridge	Date	l	s		
Hebei	Zhaoxian[1]	Yongtong Q.[3]	AD 1130	25.9	3.0	0.23	224
Hebei	Zhaoxian[1]	Jimei Q.[4] (two arches)	Probably 12th century AD	c. 8	c. 11.7	0.42	38
Hebei	Jingxing[5]	Loudian Q.[6,a]	Sui, c. AD 615	c. 18	c. 3	0.32	108
Hebei	Luancheng[7]	Lingkong Q.[8]	J/Jin, c. AD 1175	c. 20	c. 3	0.30	133
Hebei	Luancheng[7]	Guding Q.[9]	Probably 12th century AD	c. 7	c. 2	0.57	24
Shanxi	Guoxian[10]	Puji Q.[11]	Probably 12th century AD	9.4	3.6	0.77	24.5
Shanxi	Jincheng[12]	Jingde Q.[13]	Probably 12th century AD	c. 18	c. 4.2	0.46	77
Shanxi	Jincheng[12]	Zhoucun Q.[14]	Probably 12th century AD	c. 8.5	c. 4	0.94	34
Guizhou	Xingyi[15]	Muqia Q.[16]	Qing, 18th century AD	c. 20.5	c. 5.7	0.56	74
Hebei	nr. Beijing	Lugou Q.[17] (eleven arches)	AD 1191	11	c. 3.8	0.69	32

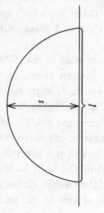

[a] The dimensions here are probably underestimated, for two spandrel arches, partly hidden by the cliff sides, can be made out in photographs.

s, sagitta (rise, radius); l, length of span (diameter or chord).

No high degree of accuracy can be claimed for these figures for many have to be taken from published scale drawings. Small differences between the figures given by responsible authorities have been averaged. But the broad lines of the general picture cannot be wrong.

1 趙縣 2 安濟橋 3 永通橋 4 濟美橋 5 井陘 6 樓殿橋 7 樂城
8 凌空橋 9 古丁橋 10 峪縣 11 普濟橋 12 晉城 13 景德橋 14 周村橋
15 興義 16 木卡橋 17 盧溝橋

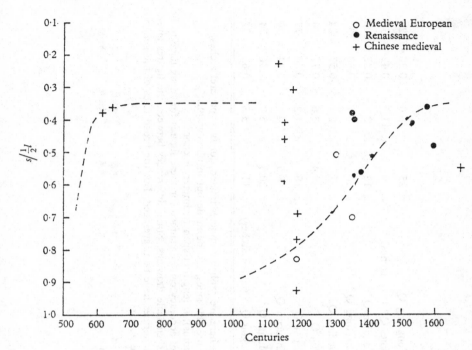

Fig. 456. Graph comparing the form of Chinese and European segmental arch bridges, the flatness ratio being plotted against a chronological scale. The degree of flatness of the 'flying' bridges of Renaissance Europe (fourteenth to sixteenth centuries AD) can thus be fully anticipated in those of early and late medieval China (seventh to twelfth centuries AD).

and *l* is the chord (in a semicircular arch this is the diameter) – see the diagram below Table 56, which gives data for a number of bridges of known dimensions in East and West and in different periods, as does Fig. 456 in graphical form.

It thus appears that in the West very slight approaches to segmentality occurred from Roman times onward, but there was only one important bridge of this type in Europe before the fourteenth century AD, the Pont St Esprit of AD 1285 (in Table 56). But its flatness ratio was only 0.51, contrasting with Florentine bridges after AD 1340 with ratios of 0.37 and 0.36. (Complete flatness is, of course, 0.00.) Meanwhile, all this had been achieved by the Sui bridges of the seventh century AD, including the masterpiece of Li Chun, the Anji Q., with its flatness ratio of 0.38, and also by the later Song bridges of the twelfth century AD, one of which, the Yongtong Q., gives the lowest ratio of any in the table (0.23). What is more, Li Chun's great bridge was not at all an isolated phenomenon, for nearly twenty others exist in various parts of China, mostly in the northern provinces, though not exclusively.

On several occasions so far it has been suggested that the segmental arch bridge was one of the inventions transmitted from China to Europe. Though almost nothing can be said about the details of the transmission, Dr Needham is not disposed to doubt the reality of the influence. The technical upsurge of the European fourteenth century AD in this particular respect points very clearly to travellers of Marco Polo's own time, though possibly the only message they brought was that the flying arch had been made by Asian men, and that it stood safe. It is, of course, even conceivable that a similar message had come through in the twelfth century AD, the time of the magnetic compass and the stern-post rudder. (See this abridgement, volume 3.)

Suspension bridges

The idea of spanning a mountain river-gorge by a suspended rope instead of a solid bridge must be very old in the history of human techniques; certainly it is very widespread. It was put into practice in the New World by many Amerindian peoples inhabiting the southern part of the continent, where cables of lianas from the tropical rain forests were used. One would naturally be inclined to assume independent invention, but facts already mentioned in connection with possible Chinese influence there in early times (see this abridgement, volume 3, pp. 153 ff.) make it necessary to be cautious about this. Among the high cultures of the Old World, such rope bridges were particularly frequent in the Himalayan region, while in the old Naki kingdom of western Yunnan they were of very primitive character, consisting of only two bamboo cables, each securely attached to high points at opposite sides of the gorge; these curved down so that their arrival points were lower than the points of departure. Man and animals passed across hanging in cradles suspended from bamboo tubes greased with yak butter; the runners were returned by means of a separate cord. In some cases the construction resembled a kind of very rudimentary cable railway. But it is the eastern part of the Tibetan massif that must surely be the focus of origin in the Old World suspension bridges, not only on account of their number, but also because so many forms, from the simplest to the most advanced pre-modern types, are found there.

The next step in the development was fixing the rope to points at more or less equal height on each side of the river, and adding arrangements which would allow travellers to cross without hanging in a cradle. One of the simplest ways in which this was done was to suspend additional ropes as handrails so that the set of three formed a V-section, the handrails being attached to the tread-rope at short intervals. This type appeared in China,

India, Burma, Borneo and Sumatra, as well as in the Celebes and Gilgit. An improvement was to add an overhead rope to the handrails, plaiting the whole together to form a continuous structure some 0.9 to 1.5 metres in diameter. Such bridges are to be found on the Assam–Tibet border, 160 kilometres north of Dibrugarh, with spans of as much as 244 metres, with a swing of 15 metres from side to side.

When we consider the techniques which could have given rise to such bridges, one which immediately comes to mind is the use of arrows with cords attached, originally devised so that the arrow could be recovered with the prey. This method appears in ancient documents such as the *Book of Odes*, and even, maybe, in late Shang bronze inscriptions of the eleventh century BC. Later they are depicted on Zhou and Han bronzes and tiles, but long before the Song (fifth century AD), the practice had grown of setting up as many as six bamboo cables horizontally a short distance apart, and then laying a proper deck of transverse planks, or lengths of bamboo, on top of them. Ropes stretched alongside on either hand to form a rail were added, and the bridge was then fit for pack animals as well as pedestrians, provided too many did not come on at one time, and the bridge was not swaying too much in the wind.

The earliest passage about such bridges found by Dr Needham is in the *History of the Former Han Dynasty* (206 BC to AD 24) begun in AD 65 and finished by AD 100. This quotes a speech of Du Qin made in 25 BC against the sending of Chinese diplomatic missions to what is now Gandhara (in modern Afghanistan), because of the extreme difficulties of the trans-Himalayan journey.

> Then comes the road through the Sanchipan gorge, 30 *li* long, where the path is only some 40 to 43 centimetres wide, on the edge of unfathomable precipices. Travellers go step by step here, clasping each other (for safety), and rope suspension (bridges) are stretched across (the chasms) from side to side. . . . Verily the difficulties and dangers of the road are indescribable.

Most geographers agree that this road (if road it could be called) was essentially the route between Suoche (or Yarkand) and Gilgit. In AD 399, Faxian, the first of the line of Chinese Buddhist pilgrims to travel to India, did not fail to describe the engineering aspects of his journey across the Pamirs, the eastern parts of the Hindu Kush:

> . . . we travelled south-westwards for 15 days. The road is difficult and broken, with steep crags and precipices in the way. The mountain-sides are simply stone walls standing straight up 2440 metres high. To look down makes one dizzy, and when one wants to move forward one is not sure of one's foothold. Below flows the Xintou He (the Indus). Men of former times bored through

the rocks here to make a way, and fixed ladders at the sides of the cliffs, seven hundred of which one has to negotiate. Then one passes fearfully across (a bridge of) suspended cables to cross the river, the sides of which are here rather less than 80 paces (about 145 metres) apart. . . .

Thus it was that the suspension bridge was virtually indispensable for communication between the people of Tibet, Afghanistan, Kashmir, Nepal, India, Assam, Burma and Thailand and those of China, where suspension bridges are still extremely common in the mountainous parts of Yunnan and Sichuan.

Already at this stage we have to note the distinction between what we are calling the rope suspension bridge and the type we know today, with its flat deck hanging level from the ropes or chains. In the first case the passengers travel along the curve, in the second they move horizontally. It seems probable that the second form arose out of the first by an excessive development of the hand-rails. Even in the bamboo-cable bridges, we begin to see a tendency for the handrails to take their origins from points higher up the banks in relation to the dipping deck than that which they occupy at midstream.

Cable suspension bridges were sometimes put to military uses in China, quite apart from the strategic importance which always tended to make them centres of combat operations. The *Wei Shu* (History of the Northern Wei Dynasty) contains an account of an ingenious submersible suspension bridge which could also be used as a boom. Constructed by a Wei general, Cui Yan-Bo who was guarding certain places on the Huai River, his object was to deny the use of its waters and its banks to the enemy, Xiao Yan, in AD 494. So,

> . . . he took the wheels of some carts, removed their rims, and cut short the spokes (to make cogwheels) so that they would engage with one another. Strips of bamboo were twisted together to make ropes, and more than ten of these cables were strung together in parallel so as to form a bridge (with cross-planks). At both ends large windlasses were set up so that the bridge was submersible at will. It could therefore neither be burnt nor cut. In this way the line of retreat of (Xiao Yan's generals), Zu Yue and the rest, was blocked, and moreover ships and boats could not get past. Thus the forces of Xiao Yan could not go to the rescue, and finally Zu Yue's brigade was all captured.

The bamboo cables of the bridges were made in the same way as those for towing ships against the river currents, but were of larger dimensions. Bamboo strips taken from the inner part of the jointed stem form a core in the rope's centre, and round them is woven a thick plaiting of bamboo strips

Fig. 457. Four of the capstans for the rail cables within the bridgehead building of the Anlan Suspension Bridge. (Photo. Joseph Needham, 1958.)

taken from the outer silica-containing layers. The plaiting is done so that the outer portion grips the core more tightly the higher the tension. Such ropes were about 5 centimetres thick, and three or more twisted together formed one bridge cable. Modern tests have shown that the straight inner strands break first, while the plaited material shows very great strength, not rupturing until a stress of 1828 kilogrammes per square centimetre is applied, that is over three times the stress that an ordinary 5 centimetre hemp rope can take.

The majority of West China suspension bridges consist of a single span, and at each end of the bridge a substantial bridge house is built on a stone foundation. Inside this house are placed two rows of stout vertical wooden columns, one for each of the side cables of the bridge (Fig. 457), and these columns act as rotating capstans for tightening the ropes. The columns are socketed into the foundations below and into beams of wood above, the whole structure being kept in place by the weight of a formidable crib of stones fitted underneath the roof. Tightening is done by hand-spikes.

The most famous bamboo suspension bridge is that at Guanxian. The Anlan Suspension Bridge, as it used to be called, has no less than eight major spans, the greatest of which is 61 metres, and its total length is over 320 metres (Fig. 458). It is 2.7 metres wide and is supported on ten bamboo

Fig. 458. General view of the Anlan Suspension Bridge at Guanxian, with its eight major spans, totalling a little over 320 metres (longest span 61 metres), taken from a hill overlooking it to the east. (Photo. Joseph Needham, 1958.)

cables, each of 16.5 centimetres diameter (Fig. 459). One of the piers is built of masonry and crowned by a decorative gate and wooden roof, but the rest are hardwood trestles (Fig. 460). The original date of construction is not known, but in view of the engineering capacity in the third century BC of Li Bing and his engineers, there seems little reason why it should not go back as far as that time. It is certainly pre-Song.

The renovation of the large Guanxian bridge puts it out of action for more than two months every year, and as maintenance is necessary for all such bridges, large or small, it was not unnatural that some more durable material should have been sought. The decisive step in the perfection of suspension bridges was undoubtedly the use of wrought iron chains, and it looks as though this invention was made in south-west China not later than the end of the sixth century AD, and quite probably in the first. The essential pre-conditions were there – traditional hanging rope cable bridges and an advanced metallurgy of iron. To accompany what is to be said, we append one additional table (Table 57) culled partly, with some difficulty, from literary sources. And to introduce the subject visually and to suggest something of the great beauty of the iron-chain suspension bridge, Fig. 461 shows the

Fig. 459. View of the Anlan Suspension Bridge looking eastwards along the easternmost span. The fastenings of the handrail cables and deck planks are clearly seen. (Photo. Joseph Needham, 1958.)

Fig. 460. A closer view of most of the Anlan Suspension Bridge from nearer water level, showing the nature of the decks and the capped trestle piers. (Photo. Joseph Needham, 1958.)

Jihong bridge across the Mekong River on the old road to Burma, seeming to match its tension with the lowering weight of the surrounding mountains.

Iron-chain suspension bridges generally have no tightening arrangements, the problem being one of anchorage only. For this purpose, massive stone abutments are built to contain the chain ends, as in the Mekong River bridge. No Chinese examples of chains covering more than one span are known, and when a pier occurs, it is built on a natural island, and the

Table 57. *Some iron-chain suspension bridges.*

Province	Location	Name of bridge	River (jiang, abbreviated to ji.)	Width (metres)	Longest span (metres)	No. of chains	Date (all dates are AD)
Yunnan	100 *li* south-west of Jingdong[1]	Lanjin T.Q.[2,a]	Lancang ji[3] (Mekong)	–	*c.* 76	20	traditionally AD 65, perhaps Sui, repaired in 1410
Yunnan	Yuanjiang[4]	Yuanjiang T.Q.[4]	Yuanjiang[4] (Hongho[5])	–	*c.* 91	–	repaired in Ming
Yunnan	350 *li* north-west of Lijiang[6]	Tacheng Guan T.Q.[7]	Jinsha ji[8] (Yangtze)	–	–	–	Sui, *c.* 595; temp. destr. in 794, captured 1252
Yunnan	85 *li* west of Lijiang[6]	Shimen Guan T.Q.[9]	Jinsha ji[8] (Yangtze)	–	–	–	Sui or early Tang, temp. destr. in 794, captured 1382
Yunnan	east of Lijiang[6]	Jingli T.Q.[10] = Zili T.Q.[11] = Jinlong T.Q.[12]	Jinsha ji[8] (Yangtze)	*c.* 3	100	18	late Qing

Table 57. (cont.)

Province	Location	Name of bridge	River (jiang, abbreviated to ji.)	Width (metres)	Longest span (metres)	No. of chains	Date (all dates are AD)
Yunnan	north of Dongchuan[13]	Junmin T.Q.[14]	Niulan ji[15]	–	–	–[b]	–
Yunnan	between Yungping[16] and Baoshan[17]	Jihong T.Q.[18] = Gongguo T.Q.[19] (?)[c]	Lancang ji[3] (Mekong)	–	69	12 (+2 rails) 1.9 cm links, 0.3 m long	ascr. San Guo,[d] iron chains by 1470
Yunnan	between Baoshan[17] and Tengyue[20]	Huiren T.Q.[21]	Nu ji[22] (Salween)	–	67 (+47)[e]	14 (+2 rails) 1.9 cm links, 0.3 m long	–
Yunnan	nr. Xiaguan[23]	–	Yangbi ji[24]	–	37	9	–
Yunnan	Binchuan[25]	Diao T.Q.[26]	–	c. 2	c. 20	2–4	–
Sichuan	near Dazhou[27] (Suiding[28])	–	Tong ji[29]	–	55	6 (+4 rails)[f]	–
Sichuan	around Emei Shan[30]	Several relatively small bridges	–	–	–	some of linked rods	–
Sichuan	Sanxia[31] (north of Jiading[32])	Sanxia T.Q.[31]	Min ji[33] (?)	–	–	–	before 1360
Sichuan	Rongjing[34] (south of Yaan[35])	–	Rongjing he[34]	–	–	4 (rods)	–
Sichuan	Lushan[36] (north of Yaan[35])	Lushan T.Q.[36]	Qingiji ji[37]	–	c. 130 (+?)[e]	–	–

Sichuan	Xiaohechang[38]	—	Fu ji[39]	—	—	7 (flat deck) (rods)	—
Sichuan	'?'[g]	—	a tributary of Min ji[33] towards Songpan[40,h]	—	23 (+3 × 23)[i]	—	—
Sichuan	Huaiyuanzhen[41] (nr. Zhongqing)[j]	Gu T.Q.[42]	—	—	—	—	Tang or pre-Tang
Sigang	Luding[43]	Luding T.Q.[43]	Dadu ho[44]	2.9	100 (formerly 110)	9 (+4 rails) 2.2 cm links, 25 cm long	1701 to 1706
Sigang	Between Yaan[35] and Dajianlu[45] (Kangding[46])[l]	—	—	—	—	(rods)	—
Sigang[k]	Changdu[47]	Diao T.Q.[48]	Jiqu he[49]	3	40	4	—
Guizhou	between Anshun[50] and Annan[51]	Guanling T.Q.[52]	Bei Pan ji[53]	11	50	30 or 36	1629[m]
Guizhou	between Shuicheng[54] and Panxian[55]	—	Hua ji[56]	—	61	4 (+6 rails)	—
Guizhou	Zhongan[57] (east of Guiyang)	Zhongan T.Q.[57]	Zhongan ji[58]	—	—	—	—
Guizhou	between Zunyi and Guiyang	—	Wu ji[59]	—	c. 60	—	—
Shaanxi	Madaoyi[60] (north of Baocheng)	Madaoyi T.Q.[60]	Bai ho[61]	—	15	6	—
Shanxi	Puji[62] (nr. Datong)	Puji T.Q.[62]	—	—	—	8	1541

Footnotes for this table are on p. 160.

Footnotes to Table 57:

^a T.Q. = *Tie Qiao* = iron-chain suspension bridge.

^b This may have been only a bamboo cable bridge.

^c The name Gongguo on modern maps may refer only to the modern suspension bridge on this road. Older maps have a Feilung[63] further up river, probably too far for it to be a synonym of the Jihong; if so, this was probably a bamboo-cable bridge.

^d This is the traditional association with Zhuge Liang.

^e Two spans.

^f The account of this bridge in the *Sui-ting Fu Chih* describes the way in which the deck planks were made to interlock by means of male and female joints secured with 'iron buttons', and also the provision of solid wooden side railings. It also mentions the windlasses used for bringing the chains to the right tension, and their fastening in position with 'stone columns'. This bridge was intended, and used, for vehicular traffic.

^g Place-name unidentifiable. Tang Huan Cheng (Ancient (and Medieval) Chinese Bridges) has Laozhunqi,[64] perhaps another bridge of the same kind.

^h The bridges in this region were regarded by the *Sui-ting Fu Chih* as the oldest of the kind.

ⁱ Four spans.

^j Characters uncertain; there is a place of this name in Guangxi.

^k This former southern province of the Tibetan marches has become Changdu Territory with loss of its eastern third to Sichuan.

^l Another iron-chain suspension bridge in this region is described by J. H. Edgar.

^m This bridge is said to have been preceded by a floating bridge made by a Daoist from Luofou Shan.

1 景東	2 蘭津鐵橋	3 瀾滄江	4 元江鐵橋	5 紅河	6 鑑江	7 塔城關鐵橋	8 金沙江	9 石門關鐵橋
10 井里鐵橋	11 梓里鐵橋	12 金龍鐵橋	13 東川	14 軍民(鐵)橋	15 牛欄江	16 永平	17 保山	18 霽虹鐵橋
19 功果鐵橋	20 勝備	21 惠人鐵橋	22 怒江	23 下關	24 漾濞江	25 黃川	26 約鐵橋	27 達州
28 綾定	29 通江	30 峨眉山	31 三峽鐵橋	32 嘉定	33 岷江	34 來蘇河	35 雅安	36 蘆山鐵橋
37 青衣江	38 小河湯	39 涪江	40 松潘	41 懷遠鎮	42 古鎮橋	43 瀘定鐵橋	44 大波河	45 打箭爐
46 康定	47 昌都	48 吊橋鎮	49 吉曲河	50 安順	51 安南	52 開嶺鐵橋	53 北盤江	54 永城
55 盤縣	56 花江	57 重安鐵橋	58 重安江	59 烏江	60 馬道驛鐵橋	61 白河	62 普濟鐵橋	63 飛龍(鐵)橋
64 老君溪								

Fig. 461. The Jihong iron-chain suspension bridge in one of the gorges of the Mekong River. With its bridgehead anchor-houses towering out of the swirling waters, the taut gentle curve (catenary) of its deck and the temples to the tutelary deities of the place visible on the left, it again affords an example of the beauty of traditional Chinese bridge building. (Photo. Popper.)

two iron-chain bridges do not form a continuous way across. The chains are always hand forged, with welded links made of bar iron 5 to 7.5 centimetres in diameter. Owing to the constant sideways movement of the bridge caused by wind in the gorges, the links near the abutments tend to wear out, and in early times replacement was not easy. Chains were not, however, the only kind of support used; at least three bridges are known which have linked iron bars. Deep in the mountains of the Sichuan–Sigang border, there is such a bridge with a span of 91 metres; its iron bars are 5.7 centimetres in diameter and 5.5 metres long joined by pin connections. It also has three intermediate stone piers, the bars passing smoothly in channels over their rounded convex tops, so that the bridge presents a gently undulating line.

Of all these bridges, perhaps the most famous, if not the most interesting technically, is the Lanjin iron-chain bridge near Jingdong in Yunnan. Local tradition and later generations believed it to have been built in the Han at the time of the emperor Ming Di (about 65 AD), though this has been challenged, but there is no doubt that it was repaired in the fifteenth century AD, and that alone antedates any iron-chain suspension bridge in Europe.

But perhaps historians have been too sceptical about the ability of the Han engineers to throw an iron-chain suspension bridge across the Mekong River. New knowledge about iron and steel technology in Han times has come about since the doubts were raised, and if it was then possible to make cast iron more than a millennium before Europeans could do so, it was surely not impossible to manufacture 76 metre lengths of chain formed of substantial wrought iron links. Since China was the focus of skill in iron and steel technology and not Tibet or Gandhāra (north-west Pakistan), we must view in a new light the undoubted existence of long-established iron-chain suspension bridges on the route to India by the beginning of the sixth century AD. Indeed, the narrative of the journey of Song Yun and Huiseng, who travelled that way in AD 519, which is given in the *Luoyang Qielan Ji* (Description of the Buddhist Temples and Monasteries at Luoyang), says:

> From the country of Bolubu (modern Gilgit in Kashmir) to the
> kingdom of Wuchang (modern north Pakistan), they use iron
> chains for bridges. These are suspended in the void, in order that
> one may cross (over the mountain chasms). If one looks downward
> no bottom can be seen, and there is nothing to grasp at in case of
> a slip, so that in an instant one may be hurled down 10 000
> fathoms. On this account travellers will not cross if the wind is
> blowing.

It seems most reasonable to suppose that the making of iron-chain suspension bridges radiated from the regions of most advanced iron technology, and it may well be that the Lanjin bridge was actually the predecessor of these bridges on the upper Indus.

Concrete, indeed decisive, evidence for the early building of such bridges comes from the time intermediate between Song Yun (AD 519) and Xuanzhuang (seventh century AD). It concerns a cluster of these bridges in north-western Yunnan in the region of Lijiang, a city at the base of a tongue of mountainous country some 96 kilometres long formed by a northern detour of the Yangtze. Above the sharp bend where the detour begins, the river was crossed by two famous iron-chain suspension bridges. The first was on the border between the Nakhi tribesfolk of the Nan Zhou kingdom and the Tibetans; here there was the Iron Bridge Town, the present village of Tacheng. This was without doubt on an important line of communication into Tibet, and apparently a special official, the Iron Bridge Commissioner, was at one time in charge of it. Kublai Khan was in these parts in AD 1252 when the bridge was captured by the Mongols, but it was old then, and must have been long since repaired, for we know that it had been destroyed in AD 794. As for the original builders of the bridge, there seems no reason for doubting the attribution to a general and military engineer of the Sui dynasty (AD 581 to 618). Other sources date it in the Kaihuang reign period (AD 581 to 600), and we know that from AD 594 to 597 Shi Wansui commanded an expeditionary force to conquer the Man tribes of Yunnan, so there was every reason for him to improve communications; indeed dynastic histories specifically mention his river crossings during the campaign.

Further down the river there was another great iron-chain suspension bridge across the Jinsha River near Shimen Guan. If it was not built by the general Shi Wansui and the military engineer Su Rong in the 590s, it must have been made not much later, for in the same Tibetan campaign of AD 794 it too was destroyed; now it no longer exists. Finally, another iron-chain suspension bridge, the Jinlong Bridge, is still very much in use (Fig. 462); it links Lijiang to the east across the now south-flowing river with the mountain town of Yongbei.

The gorge at Gongguoqiao has had an iron-chain suspension bridge since about AD 1470, hence the name Jihong Bridge or 'rainbow in a clear sky'. Provincial history avers that holes in the rocks for cables or chains were made under the orders of Zhuge Liang, the great Captain-General of Shu, during his conquest of Yunnan in AD 225 to 227. Though confirmatory evidence is lacking, there is nothing at all impossible in the tradition. Certainly, the historical geographer Gu Zuyu (*flourished* AD 1667) recorded that the engineer Wang Huai (*flourished* AD 1488) converted the bridge, linking iron chains together and adding cross-planks over which people could walk as if on level ground. But this was a period of many conversions, for we know of yet another engineer, Zhao Qiong, who replaced at least one of the cable bridges over the Longchuan River in northern Yunnan by an iron-chain bridge at about the same time.

Accounts of military operations reveal a close connection between suspension bridges, pontoon bridge cables and defensive harbour booms; all form a single technological complex. Interesting examples of the use of iron chains may now be added to the case of the submersible bamboo cable bridge-boom of the late fifteenth century AD already described (p. 153). There were many ways of using suspended chains besides capturing existing bridges or denying their use to hostile forces. The eleventh century AD *Wu Dai Shi Ji* (Historical Records of the Five Dynasties Period) tells of a battle in AD 928:

> In the fourth year of the Bailong reign period the army of Chu attacked Fengzhou (on the West River near the western border of modern Guangdong) with numerous ships, and defeated the defending forces on the He River. (Liu) Yan was alarmed . . . he sent his general Su Zhang with a 'Magic Crossbow Division' of 3000 men to the relief of Fengzhou. (Su) Zhang took a pair of

Fig. 462. A traditional iron-chain suspension bridge of the Qing period, the Jinlong bridge linking Lijiang with Yongbei across the Yangtze in Yunnan. Eighteen chains carry the road across a single span of 100 metres. The great river here is running at an altitude of 1402 metres, and the masonry of the bridgeheads is built up, as can be seen, to withstand an annual rise and fall of some 18 to 21 metres or more. (Photo. from Rock (1947).)

iron chains and sank them deeply in the He River, with very large winches (or capstans, lit. wheels) on each bank (to tighten them), and tamped earth redoubts to conceal (the winches and their crews). Then he invited battle with light boats which, pretending defeat, fled, hotly pursued by the men of Chu. At the right moment (Su) Zhang set in motion the great wheels, which hauled up the booms and cut off the Chu ships, exposing them to the cross-fire of powerful arcuballistae [crossbows] set up on each bank, so that hardly a man of the Chu forces escaped to tell the tale.

Chinese military engineers must have had recourse many times through the centuries to these techniques. For again, in AD 1371, technicians of Shu flung boldly across one of the greatest of the Yangtze gorges three cable suspension bridges commanding as many iron-chain booms, and equipped with bomb-throwing trebuchets (stone hurling catapults) and all kinds of firearms.

Just as the arch bridge found its Boswell in Xie Gantang, so the iron-chain suspension bridge also rose, in one case, to the dignity of a book. This concerned the bridge at Guanling over the Northern Panjiang, in south-western Guizhou. The *Tie Qiao Zhi Shu* (Record of the Iron Suspension Bridge) was written by Zhu Xieyuan and was printed in AD 1665, though it concerned the erection of the bridge which had taken place already in 1629. The author's father, Zhu Jiamin, had been the prefect under whose auspices the bridge had been built, so Zhu himself had access to all the official documents. His book was illustrated by a panoramic drawing of the structure and its approaches, part of which is shown in Fig. 463. From the famous explorer traveller, Xu Xiake, who visited the bridge in 1638, not long after it was finished, we learn that its span was 45 metres and that it had a floor of wooden planks, each 20 centimetres thick and more than 2.4 metres long. He also tells us that prior to its construction ferries had been used but these were often in danger of capsizing, and that an attempt at a stone bridge had failed. He also remarks on its strength, and comments that 'daily hundreds of oxen and horses with heavy loads pass over it'. Its sides were protected by 'a high iron railing woven with smaller chains', while on each bank these railing chains are clenched in the mouths of two stone lions, each 0.9 to 1.2 metres high.

Zhu's remark about the railings is of technical interest because since the railings were also made of chains, these would take a portion of the weight, and thus invited a transition to a flat deck suspended entirely by a hanging chain. Though partly destroyed in AD 1644 but repaired in 1660 and many times later, a modern steel structure did not replace the original until 1939.

Fig. 463. An Illustration from the *Record of the Iron Suspension Bridge* of AD 1629 by Zhu Chaoyuan, showing the Guanling bridge in the gorge of the northern Pan Jiang, between Anshun and Annan in south-western Guizhou. Thirty to thirty-six chains bridged the span of some 50 metres. Erected in AD 1629, it was replaced in 1943 by a steel-chain suspension bridge at a broader but easier place nearly 1 kilometre downstream. Zhu's drawing has many interesting features. Under the bridge we read: 'The water is so deep here it has no bottom.' To the right, in the background, there are a number of temples, including a pagoda and a library for the *sūtras*. To the left there is a masonry embankment against which break the 'hundred foot waves.' In the foreground on the right there is a Buddhist statue, and on the left the 'stone of weeping', a monument to those who perished in the crossing before the bridge was built.

It is possible that the work of the Guizhou bridge builders was brought to the notice of Europeans only a few decades after Xu's description. For in the map of this province of Blaeu's great *Atlas* (entitled *Novus Atlas Sinensis*) the Jesuit, Martin Martini, in AD 1665 marked an iron-chain bridge so far unidentifiable but probably somewhere near Guanling. Of course, Martini may not have seen this himself during his travels in China, but the information could easily have reached him from his colleagues. An earlier representation of such a bridge than that in Fig. 463 is given by Wang Zhenpeng, who flourished in the fourteenth century AD. In his painting made between 1312 and 1320, giving his imaginative construction of a Tang palace, an iron-chain bridge is shown slung across the mouth of a huge cavern.

There is still the question of the metamorphosis of these traditional iron-chain suspension bridges to the flat-deck type. In the 1880s, W. Gill who was journeying in China, found a flat-deck bridge in northern Sichuan, and Fugl-Meyer, though not personally acquainted with the region, maintained that in the Tibetan and Himalayan bridges the load was generally carried by two slack main cables with a flat deck slung below. Another such bridge crosses the Brahmaputra River with a span of 137 metres; it is known to have a flat deck from a sketch made in 1878. It had a tower at each end but the deck was only wide enough for pedestrians. The date of construction of the bridge is given as AD 1420, and the engineer is said to have been Thaṅ-ston-rgyal-po, who, tradition has it, lived from AD 1361 to 1485; his Tibetan title was 'Builder of Iron Bridges'.

It would also be helpful to know the date of the first iron-chain bridge in Bod-Yul. The one mention of this bridge seems to be in the Tibetan chronicle found at Dunhuang, which relates to the year AD 762, and says:

> As the Chinese government had collapsed, it was not the right
> time for presenting tax-silk and maps of the country. (On the
> contrary) Zaṅ-rgyal-zigs and Zan-stoṅ-rtsaṅ, having crossed the
> iron bridge at Bum-liṅ, invested with their forces many Chinese
> cities . . . which fell into their hands . . .

This is but a gleam in the darkness, however, for we do not know how old the bridge was at this time, nor whether it was built by the Tibetans or the Chinese. Nor do we know its exact location, though from the context it was probably somewhere on the upper Yellow River south of Xining, and doubt-less commanded the approaches to Gansu.

As far as the history of suspension bridges in Europe is concerned, not the sixth century AD but the sixteenth brought the first Western specification for a suspension bridge. In 1595, Faustus Verantius proposed two towers, a flat deck and a system of linked rods or inverted brackets, formed of chains of

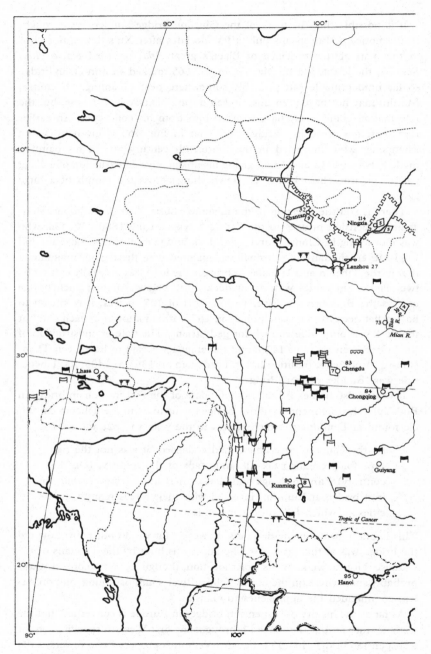

Fig. 464. Map to illustrate the engineering works of ancient and medieval China.

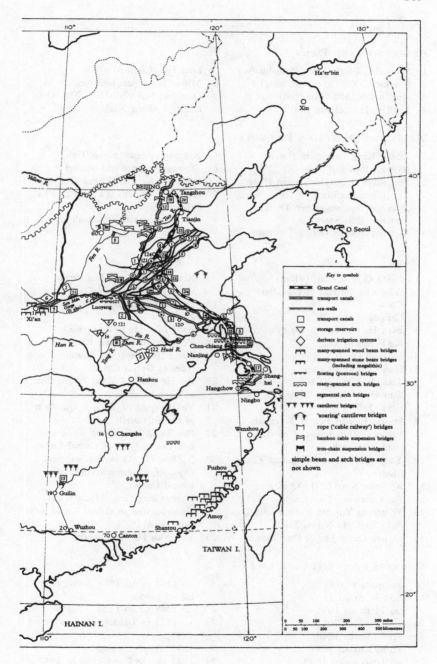

Key to symbols

	Grand Canal
	transport canals
	sea-walls
	transport canals
	storage reservoirs
	derivate irrigation systems
	many-spanned wood beam bridges
	many-spanned stone beam bridges (including megalithic)
	floating (pontoon) bridges
	many-spanned arch bridges
	segmental arch bridges
	cantilever bridges
	'soaring' cantilever bridges
	rope ('cable railway') bridges
	bamboo cable suspension bridges
	iron-chain suspension bridges

simple beam and arch bridges are not shown

Key to Fig. 464:

RESERVOIRS AND DAMS

1 Shao Bei; Peony Dam (Sunshu Ao)
2 She Gong Bei (Shen Zhuliang)
3 Qianlu Bei (Shao Xinchen)
4 Xinfeng Ho (Zhang Kai)

5 Lian Hu (Chen Min)
6 Mulan Bei (Qian Siniang)
   Qiantang Sea-Wall (Hua Xin, Qian Liu and Zhang Xia)

WEIRS AND DIVERSION PROJECTS

1 Zhang River Irrigation System (Ximen Bao and Shih Qi)
2 Fen River Irrigation System (Pan Xi) (unsuccessful)
3 Shouxian System (Chen Deng) (remains still extant)
4 Meixian System (Kong Tianjian)

5 Ningxia System (Meng Tian)
6 Zhengguo Canal and Weibei Irrigation System (Zheng Guo)
7 Guanxian System (Li Bing and Li Erlang)
8 Kunming System (Shansiding)
9 Shandan System

CANALS

1 Hong Gou = Bian (=Ban) (Zhou to the end of Nan Bei Chao)
2 Langdang Qu
3 Bian He = Tong Ji Qu (Yuwen Kai) (Sui to Yuan)
4 Guangji Qu (Qi Huan)
5 Han Gou (Fu Chai)
6 Tao Qu (Fu Chai)
7 Cao Qu = Guang Tong Qu
8 Chu Qu
9 Hutuo R. and Fen R. R. canals (unsuccessful, replaced by cart-road portages)
10 Yang Qu (Wang Liang and Zhang Shun)
11 Kaiyuan Xin He (Li Qiwu)
12 Bai Gou etc. (Cao Cao)
13 Shanyang Yundao = Shanyang Du = Li Yun He (Chen Min)
14 Donga Qu = Jing Ji Du (Xun

Xian) (incorporated in Grand Canal)
15 Ling Qu (Shi Lu)
16 Zhan Qu
17 Jiang Nan He (Yuwen Kai)
18 Yong Ji Qu (Yuwen Kai)
19 Tong Hui He (Guo Shoujing) (part of Grand Canal)
20 Bai He (part of Yong Ji Qu, then of Grand Canal)
21 Yu He (part of Yong Ji Qu, then of Grand Canal)
22 Hui Tong He (Zhang Kongsun and Loqsi) (part of Grand Canal)
23 Ji Zhou He (Guo Shoujing and Oqruqči) (summit section of Grand Canal)
24 Huan Gong Gou (Huan Wen) (incorporated in Grand Canal with extensions)
25 Yun Yan He (a network)

COURSES OF THE YELLOW RIVER

0 antiquity to 602 BC
1 602 BC to AD 11
2 AD 11 to AD 1048
2a AD 11 to AD 70
2b AD 70 to AD 1048
3 AD 893 to AD 1099
3a AD 1060 to AD 1099
4 AD 1048 to AD 1194

5 AD 1194 to AD 1288 (some flow till AD 1495)
6 AD 1288 to AD 1324
7 AD 1324 to 1855 (entirely after AD 1495)
8 1855 to present date
8a 1887 to 1889 and 1938 to 1947

N.B. Numbers indicating towns and cities are the same as those in Table 51 (pp. 8–12), and reference there will identify them.

bars with an eye at each end. The fact that bridges of rods had already been used in China is rather remarkable in this connection. But there is no proof that Verantius gathered anything from information which Portuguese travellers of the early part of the century brought back with them; any suspicions we may have must depend on the relative closeness of the dates. At all events, Verantius did not actually construct such a bridge. Martin Martini referred to the Jingdong iron-chain suspension bridge in AD 1655, and it was later referred to in 1667, 1725 and 1735. According to the engineer Robert Stevenson, the earliest iron-chain suspension bridge in Europe was the Winch Bridge over the Tees (AD 1741) in northern England, and, perhaps significantly, this was of the hanging chain, not the flat-deck type. The first suspension bridge capable of carrying vehicles was not built until 1809 in Massachusetts; it had a span of 74 metres. This was followed in 1819–26 by Telford's Menai Straits Bridge, with a span of almost 177 metres. After that, such bridges became commonplace, and Dr Needham feels driven to the conclusion that there must have been, in this whole succession of events, a real series of stimulations from the Chinese iron-chain suspension bridges to the engineers of the Renaissance and later Europe, even though not all the stages have been elucidated.

Geographical distribution of types of bridges

If we glance at the map (Fig. 464), on which the locations of many bridges we have discussed have been marked, we can see that Fugl-Meyer was broadly right in dividing China into three regions from the point of view of bridge engineering. In the northern zone, down to the northern parts of Zhejiang, Jiangxi and Henan, arch building was predominant, beam bridges being reserved for minor or decorative structures. This was the zone of the segmental arches and long multiple-span arch bridges. In the western zone, on the other hand, which includes besides Yunnan, Sichuan and Guizhou, the two former provinces of the Tibetan marshes (Sigang and Qinghai) and also Gansu and part of Shanxi, all bridges of importance were cantilever structures or suspension bridges. Beam bridges are rarely found there. In the southern zone, centring on Fujian but including Guangdong and Guangxi, beam bridges are the commonest type, culminating in the megalithic giants of the Fujianese coast. Here, arch bridges are only for minor or decorative purposes, and cantilever and suspension bridges are never seen.

While topographical reasons may account for this distribution, it does not seem impossible that it might have some connection with those local cultures to whose fusion Chinese society owes its existence. But the nature of the terrain and the materials available must have played a part at least equal to invention and its subsequent stylisation within particular ethnic or social groups.

5

Hydraulic engineering (I), control, construction and maintenance of waterways

If there was one feature of China which impressed the early modern European travellers more than any other, it was the great abundance of waterworks and canals. In AD 1696, Louis Lecomte wrote:

> Tho' China were not of it self so fruitful a Country as I have represented it, the Canals which are cut thro' it, were alone sufficient to make it so. But besides their great usefulness in that, and the way of Trade, they add also much Beauty to it. They are generally of a clear, deep, and running Water, that glides so softly, that it can scarce be perceived. There is one usually in every Province, which is to it instead of a Road, and runs between two Banks, built up with flat course Marble Stones, bound together by others which are let into them, in the same manner as we use to fasten our strong wooden Boxes at the Corners.
>
> So little Care was taken, during the Wars [the Manchu conquest half a century previously], to preserve Works of Public Use, that this, tho' one of the Noblest of the Empire, was spoiled in several places, which is a great pity; for they are of no little use, both to keep in the Waters of these Canals, and for those to walk on who drag the Boats along. Besides these Cawseys they have the conveniency of a great many Bridges for the Communication of the opposite Shoars; some are of three, some five, and some seven, Arches, the middlemost being always extraordinary high, that the Boats may go through without putting down their Masts. These Arches are built with large pieces of Stone of Marble, and very well framed, the Supporters well fitted, and the Piles so small that one would think them at a distance to hang in the Air. These are frequently met with, not being far asunder, and the Canal being strait, as they usually are, it makes a Prospective at once stately and agreeable.

This great Canal runs out into smaller ones on either side, which are again subdivided into small Rivulets, that end at some great Town or Village. Sometimes they discharge themselves into some Lake or great Pond, out of which all the adjacent Country is watered. So that these clear and plentiful Streams, embellished by so many fine Bridges, bounded by such neat and convenient Banks, equally distributed into such vast Plains, covered with a numberless multitude of Boats and Barges, and crowned (if I may use the expression) with a prodigious number of Towns and Cities, whose Ditches it fills, and whose Streets it forms, does at once make that Country the most Fruitful and the most Beautiful in the World.

Lecomte was thus full of admiration for the hydraulic engineers of China. He realised that their work went back even into the legends, for he went on:

The Chinese say their Country was formerly totally overflowed, and that by main Labour they drained the Water by cutting it a Way through these useful Canals. If this be true I cannot enough admire at once the Boldness and Industry of their Workmen who have thus made great Artificial Rivers, and a kind of Sea, and as it were created the most Fertile Plains in the World.

Lecomte was also quite clear about the dual function of the waterways, for transportation and for irrigation. Thus he says that the 'Great Canal' was

necessary for the Transportation of Grain and Stuffs, which they fetch from the Southern Provinces to Pekin. There are, if we may give credit to the Chineses, a Thousand Barks, from Eighty to a Hundred Tun, that make the Voyage once a year, all of them Freighted for the Emperor, without counting those of particular Persons, whose number is infinite. When these prodigious Fleets set out, one would think they carry the Tribute of all the Kingdoms of the East, and that one of these Voyages alone was capable of supplying all Tartary wherewithal to subsist for several years; yet for all that Pekin alone hath the benefit of it; and it would be as good as nothing, did not the Province contribute besides to the Maintenance of the Inhabitants of the vast City.

The Chineses are not only content to make Channels for the Convenience of Travellers, but they do also dig many others to catch the Rain-Water, wherewith they water the Fields in time of Drought, more especially in the Northern Provinces. During the whole Summer, you may see your Country People busied in raising this Water into abundance of small Ditches, which they contrive

across the Fields. In other places they contrive great Reservatories
of Turf, whose Bottom is raised above the Level of the Ground
about it, to serve them in case of Necessity. Besides that, they
have everywhere in Chensi and Chansi, for want of Rain, certain
Pits from Twenty to an Hundred feet deep, from which they draw
Water by an incredible Toil. Now if by chance they meet with a
Spring of Water, it is worth observing how cunningly they husband
it; they Sustain it by Banks in the highest places; they turn it here
and there an Hundred different ways, that all the Country may
reap the benefit of it; they divide it, by drawing it by degrees,
according as every one hath occasion for it, insomuch that a small
Rivulet, well managed, does sometimes produce the Fertility of a
whole Province.

Even allowing for a little enthusiasm, Lecomte was perfectly right that the
Chinese people have been outstanding among the nations of the world in
their control and use of water.

The purpose of this chapter will be to examine more closely the nature
of their achievements and the engineering techniques which they developed
in the process. But to begin with we must sketch a broad view of the
problems the Chinese people had to face and the solutions they adopted.
There was the climate, with its special features of rainfall. There was the
topography, and the peculiarities of the great river-systems which formed
the given framework for human enterprise, involving the first of the great
requirements, protection from floods. The second great requirement, that of
irrigation systems, was dictated partly by the nature of the loess soil of the
upper Yellow River basin, and partly by the widespread adoption of wet rice
agriculture (Fig. 465). The Chinese may constitute one-fifth of the world's
population, but their irrigated land is one-third of the world's total, over 40
million hectares out of some 120 million hectares. Thirdly, the more centralised the feudal-bureaucratic State became, the more essential was the
construction of waterways along which tax grain could be transported; and
this led naturally to a fourth factor, namely the military aspects of defence.
Centralised granaries and arsenals could furnish army supplies at need,
while canals were themselves an important obstacle to the penetration of the
Chinese farming civilisation area by cultures of nomadic type. All this is
contained in the classical Chinese term for hydraulic engineering, still used
today, *shui li* (水利), 'benefit of water'.

To complete this introduction, a few words are necessary about what
hydraulic engineers through the centuries could hope to do with water. The
fundamental unit is of course the river valley, and the harnessing of a river
can take place in one or more of the following ways:

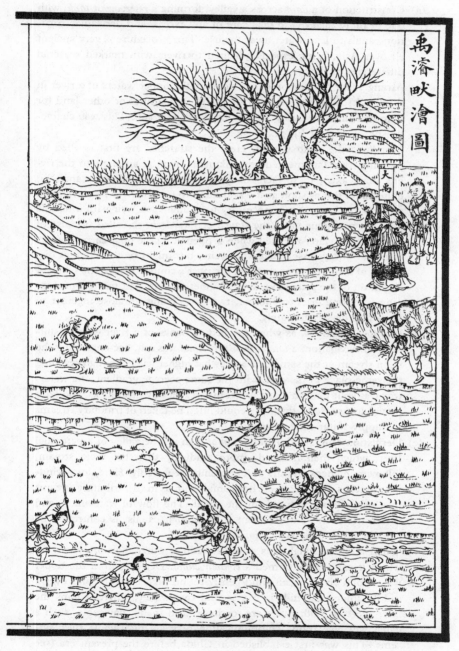

Fig. 465. A late Qing representation of Yu the Great exhorting irrigation workers, from *The* Historical Classic *with Illustrations* of AD 1905. The caption quotes the words: 'I caused the channels and canals to be dug and deepened . . .'.

(a) Construction of a *dam* across a valley, forming a *reservoir* or *tank*, with one or more *spillways* to take care of any excess, and derivative irrigation canals to lead water away for use. This procedure is very ancient all over Asia, being especially suited to rivers with marked seasonal differences of flow.

(b) Arrangement of *retention basins* whereby the upper waters of a river in the flood season are made to submerge agricultural or other land for a limited time, restoring fertility by depositing soil. This was characteristic of ancient Egypt.

(c) *Canalising* the main stream. Here the stretches are first levelled by building *weirs* (submerged embankments often at an angle to the river's main axis); these facilitate getting water to irrigation canals on either side. But as soon as it was desired to use the river for navigation at the same time, it was necessary to couple each weir with a double *slipway* or a single gate (the *stanch* or *flash-lock*, so-called because of the 'flash' of water released when it opens). These boat-carrying ways were not always associated with weirs, but might be set alone at distances of a kilometre or so along the river's course. Lastly, the Chinese invention of the *pound-lock* assured quiet and efficient transfer of shipping from one level to another, two gates being set close together to open and close alternately.

(d) In other situations, as when the bed or banks of a river were unsuitable for navigation, a *lateral transport canal* was cut, accompanying the main stream at the same or a somewhat higher level. Though reaching the same place in the end, with the same average gradient, the waterway was split up into many horizontal reaches by flash-locks, and later pound-locks. Here advantage could often be taken of tributary streams to supply water to the canal.

(e) Derivation of a *lateral irrigation canal* high up the valley of a perennial river, which then descends more gradually than the main stream, following higher contours and branching repeatedly. Here, long winding stretches may be almost horizontal, and, as in (d), tributary streams may usefully be captured for the canal. The flow of water in it, and the distribution of water from it by branch canals, is controlled by suitably adjustable gates (*sluices*).

(f) When a lateral canal ends in a watercourse other than the one from which it came, it is termed a *contour canal*. Deriving from the upper waters of a river, it can wind gently round high contours and over a saddle among the hills into a second river valley, where it joins the river there. If both rivers are, or have been made, navigable, such a canal will give traffic communication between two whole river systems. This was first established in China before the present era (see pp. 212 ff.). Similarly, a contour canal can irrigate otherwise unusable land outside the river valley in which it originated. It may also go to feed a reservoir in another valley, and may thus combine the waters of perennial river systems as well as those which become inundated.

(g) Connection between two river systems can also be achieved by the *summit level canal*, which scales the contours directly on each side of a range. If the watershed was a very low one, double slipways may be sufficient for handling the traffic. But steeper gradients could not be attempted until it became possible to build flash-lock gates with some success; and such routes did not become fully practicable until the invention of the pound-lock. Such a lock could then raise or lower one or two vessels at the same time some 3 or 6 metres, thus avoiding haulage over slopes or against rushing discharges. There was, of course, always the difficulty of ensuring an adequate supply of water for the summit levels, but many ingenious ways of doing this were devised. Again all this was first accomplished in China (see pp. 222 and 249).

Half a millennium before our era, Chinese rulers and engineers were well aware that water transport was highly efficient mechanically. Not until the coming of the railways and motor vehicles was there any other satisfactory way of carrying heavy loads from place to place. This may at once be seen from comparative loads (given in tonnes) carried or drawn by a single horse:

pack horse		0.127
horse harnessed to wagon	– 'soft road'	0.635
	– macadamised road	2.0
	– iron rails	8.0
horse harnessed to barge	– river	30.0
	– canal	50.0

In what follows, we shall trace the epic development of the transport canal in China, sometimes evoked by the need for military supplies, but more often inspired by the nature of the fiscal system, which sought to concentrate the products of the people's labour at the bureaucratic nerve centre of a vast terrestrial empire.

PROBLEMS AND SOLUTIONS

The most fundamental factor with which those who controlled the waterways had to contend was the Chinese climate. Something about this was said in volume 2 of this abridgement (pp. 222 ff.), but here we are concerned with it only so far as precipitation of rain determined the size and character of the natural rivers.

The rainfall (Fig. 466) has a highly seasonal distribution, around 80 per

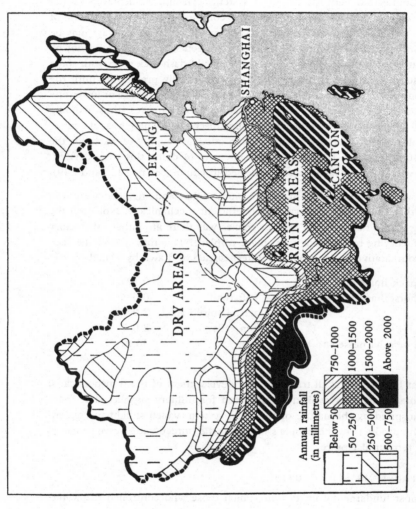

Fig. 466. Rainfall map of China (after Luo Gaifu, modified by Cedric Dover). The dark line separating the two great areas indicates very nearly the effective northern limit of wet rice cultivation, though this is to some extent a customary division since rice will mature in all parts of China. The Tibetan region to the west (left) is too high for such agriculture. Extremes of rainfall here shown range from 50 to 2000 millimetres per annum.

Annual rainfall
(in millimetres)

Below 50
50–250
250–500
500–750
750–1000
1000–1500
1500–2000
Above 2000

cent of it occurring in three summer months. At the same time, the prevailing wind changes direction; this is the phenomenon of the monsoon. In winter the air masses over inner Asia are cooled and tend to sink, expelling the moist oceanic air from China, and producing dry, cold, north-westerly winds. In summer the reverse takes place; the central air masses are warmed and rise, so that conventional circulation occurs, bringing in south-easterly winds carrying humid air from the south-eastern seas. As an example, near Liuzhou in the southern province of Guangxi, the four summer months are much wetter than the rest, and analysis of the rainfall figures implies that the watercourses will often be nearly dry for most of the year, and that flooding is sudden. It was necessary, therefore, to build works which would withstand torrents much greater than the winter flow. And though, on the whole, the total rainfall to be expected was greater in China than Europe, its intensely seasonal character set a tremendous problem – the building of enough reservoirs to prevent the water from running uselessly away.

Still greater difficulties were caused because the annual monsoon climate shows an annual fluctuation of rainfall much greater than in other parts of the world. Works were therefore necessary on a scale sufficient to take care of even the most exceptional years, and naturally it was a long time before this ideal was achieved. In Europe the ratio of the wettest to the driest years hardly ever exceeds 2 for any given place; at Shanghai the ratio for a 50-year period was 2.24. For individual months it might be much higher; for example, in 1886 there was only 3 millimetres of rain in July compared with 306 millimetres in 1903.

The rainfall dissipates itself throughout the great rivers of China, and their physical characteristics have naturally determined the conditions of life of the people and the greatest works of protection and control. Of the four river systems, the Yellow River (Huang He), was the most difficult to master and was important in earliest Chinese history, followed by the Huai River and the Yangtze, setting serious problems of their own, and finally the Canton river system. Though the central and southern rivers flow through fertile rice growing regions and became connected by an elaborate system of canals, Chinese hydraulic engineering served its apprenticeship in the hard school of the Yellow River valleys, tackling problems which even modern technology has not yet solved.

The Yellow River rises in the relatively dry plateau of north-eastern Tibet and descends quickly eastwards through a vast area covered by easily eroded loess soil. About halfway along its course it comes up against the Shanxi massif and turns south, continuing on in that direction until it turns abruptly eastwards at Tongguan, after which it emerges from the mountains toward Kaifeng and pours over the alluvial North China Plain. The river's upper basin, the cradle of Chinese culture, is covered with loess, except for the

Ordos Desert, through which it runs at the top of its great bend. Similarly the low-lying plain east of the mountains is all alluvial loess. As it comes down from the mountains, it arrives at another mountain range on the peninsula of Shandong and must go either north or south of it on its way to the Yellow Sea. It has in fact taken both these courses at different times.

From its source to the sea the Yellow River covers some 4650 kilometres, with a drainage area of almost 770 000 square kilometres. But this area does not now include the alluvial plane through which it runs for 7000 kilometres after it emerges from the mountains; and more than 155 000 square kilometres of this land cannot be included in the watershed figure since the river has risen above the surrounding plain, from which it is cut off by dykes. Indeed, the bed of this 'elevated river' constitutes a ridge separating two drainage basins. In early times, however, the river must have wandered about all over this plain and accepted drainage from it. Because of the great fluctuations in rainfall, there are tremendous variations in the river's discharge. In the dry season it is similar to that of the Thames, but at its greatest may be of the order of over 28 000 cubic metres per second, a figure only equalled by the Yangtze. However, the problem with the Yellow River is much more serious, for although the Yangtze is subject to devastating floods, it is at least surrounded throughout its length by higher ground forming a narrower channel. But what makes the Yellow River almost unique in the world is the prodigious quantity of silt which it carries down, perhaps at least as much as one million tonnes per annum. In the light of these figures, one can begin to understand the magnitude of the problems which faced civil engineers when, from the Qin and Han onwards, State centralisation began to give them the possibility of concentrating large masses of manpower on the construction of works for flood protection.

The Yangtze is much longer than the Yellow River, attaining a length of some 5550 kilometres, and drains a much larger area, about 3 160 000 square kilometres. Though the water is generally coffee-coloured, it carries only about 0.2 per cent silt compared with the 10 per cent to 11 per cent of the Yellow River's main stream, and the 30 per cent or so of the tributaries in the loess area. After coming from the Tibetan mountains and passing through Sichuan, the Yangtze plunges through famous gorges at an average gradient of about 19 centimetres per kilometre. Rapids due to rocky dykes make it dangerous for navigation at low water. At one point there is a difference of 60 metres between high and low water, and even at Chongqing it averages 30 metres; in the gorges some of the pools have a depth of 60 metres. Between Yichang and Nanjing there are wide areas on each side of the river liable to flooding, and after Zhenjiang there is no more safe high ground before the sea is reached, but the total area which the river can inundate is smaller than that at the mercy of the Yellow River. This is good, for the Yangtze valley is the most populated in the world, containing more than one-third of China's total population and producing 70 per cent of its rice.

The drainage area of the Huai River (more than 274 000 square kilometres) has been included in the figure for the Yangtze, since, for many centuries after its own exits to the sea on the coast round Shandong were blocked, it has delivered its waters into that river; the north–south line of communication thus formed became part of the Grand Canal. In comparison with the two great rivers, the Huai and the West River of Guangdong (some 1930 kilometres long and draining about 696 700 square kilometres) are not considerable, yet the former has set many intractable problems, and only in modern times have those of the Huai valley been firmly taken in hand.

The paramount social importance of floods and their prevention is seen especially well in a section of text from the third or early second century BC in the *Guan Zi* (Book of Master Guan). Duke Huan is questioning Guan Zhong on the best location for the State capital, and the minister mentions the 'five harmful influences', flood, drought, unseasonable wind, fog, hail and frost – but of these floods are the worst. For this reason, he suggested that Water Conservancy Offices should be established in each district, and staffed by those 'who are experienced in the ways of water'. For the area on each side of the river one man from the staff should be selected as chief hydraulic engineer, and so in charge of inspecting the waterways, the walls of cities and their suburbs, the dykes and rivers, canals and pools, government buildings and cottages, and to supply those who carry out the repair work with just enough men. Indeed, details were laid down about not only mustering sufficient *corveé* labour (both men and women), but also suitable strengths of armoured weapon-bearing soldiers for the repair and maintenance of dykes, to say nothing about the specification for the number of spades, baskets, earth tampers, planks and carts for each detatchment. Moreover, the regulations said that summer and autumn, being the times of hoeing and harvest, should never be used for public works, but that the winter was suitable for inspection and the accumulation of stores, while most of the work must be done during the spring agricultural lull when water levels are low.

Chinese agrarian society was from the beginning based on intensive agriculture, and, for this to succeed, irrigation works both small and large were as necessary in its culture as coal mining and iron casting were for a later age. In the north-western loess region, the problem was primarily one of leading off contour canals from natural watercourses, making them fall through a more gentle gradient, and distributing the water to the fields. In the Yangtze valley and the valleys of Guangdong, the problem was the drainage of the fertile by swampy land and the maintenance of a system of canals for this purpose. In the basins of the Huai and the Yellow River, the main necessities were the construction of works stout enough to withstand or delay the greatest volumes of flood water and dams or reservoirs to retain the water of the rainy season and release it gradually.

Modern studies of loess show that this spongy soil is highly porous, and capillary action within it helps the rise of mineral elements from the subsoil to the roots of plants. Rich in potassium, phosphorous and lime, it needs only abundant water with the addition of organic manure to show great fertility. And what is true of the loess soil already there is true also of the loess carried down towards the sea in the form of silt by the great rivers. Chinese peasants and officials appointed long ago in the first millennium BC appreciated the importance of this silt as a fertiliser, and faced very consciously the complex problems relating to its control. They used ever higher dykes to prevent it obstructing the course of the river, and distributed or retained it with sluice-gates and other devices. At an early date, too, there was a recognition of the relation of this silt flow to the deforestation and denudation of the mountains. Of course, such questions were bound up with social problems. In the north, peasants could not easily be dissuaded from settling on the rich land within the great dykes which was only occasionally flooded. In the south, landowners and rich farmers tended to encroach upon the lands recovered from drained swamps and lake bottoms – land nominally belonging to the State but unclaimed by any individuals – with the result that reservoir space available in flood times was greatly diminished. Nevertheless, irrigation and the fertilising properties of silt gave the clue to the achievement of China maintaining 'permanent agriculture' over the centuries.

Hydraulic engineering was doubly demanded in China. While in the north, irrigation projects were suggested by the nature of the loess soil, even though the main crops were dry – wheat, millet, etc. – in the south, abundant water was indispensable for wet rice cultivation. All the Chinese agricultural treatises emphasise the interest which the farmer should take in the supply of water to his rice fields. The regulation of water is indeed the utterly essential prerequisite of wet rice farming, and many Chinese drawings show work being done on dykes and channels, a constant occupation of countrymen in off seasons.

Throughout Chinese history, the taxes essential for unified State power were collected in kind, and most of this payment was in grain. Such grain tribute was the fundamental source of supply for the imperial clan, the central bureaucracy and the army, with its headquarters at the capital. Indeed, for the whole period of bureaucratic feudalism, the government generally considered the interests of grain transport as overriding those of irrigation and flood control. One can see this particularly well in the case of the Grand Canal, which crossed the Huai valley at right angles, and often interfered with the rational conservation of the Huai waters.

The importance of water transport was by no means limited to times of peace; even more in times of disturbance and war, whether civil or external, the command of the waterways as supply routes was of inestimable value. For instance, after the fall of the Qin dynasty, Liu Bang, founder of the Han dynasty, owed much of his success in defeating his powerful rival, Xiang Yu, to his control of the Wei valley. From that secure base, it was possible to use water transport to send supplies of grain to the Han armies.

The military significance of canal building was not, however, confined to the problems of supply. Networks of waterways and ditches formed a kind of defence in depth which gave almost insuperable difficulties to nomadic armies chiefly composed of cavalry. This comes out particularly in the wars between the Liao (Khitan Tartars) and the Song. The Liao often lost time while ditches were filled in and bridged, or while barges were built for ferrying, and the walled cities of China with their moats, connected by good canals but inferior roads, constituted an almost ideal series of strong points scattered over the country which it was difficult to reduce. This was very different from the steppe country, for which armies of horsemen were well suited.

SILT AND SCOUR

We turn now to more detailed questions. The fertilising effects of silt on salty and alkaline land were appreciated as early as 246 BC, when the Zhengguo Irrigation Canal was completed, the first of a long series of irrigation projects in the Wei valley. In the early days, silt was generally regarded as wholly advantageous, but as time went on it was realised that too much might be a disaster. If it originates from topsoil and is not deposited too thickly it is good, but it may damage the land severely when it comes from eroded subsoil.

The dangers of silt can be recognised in many reports, for example from the tenth century AD, but already in Wang Mang's time (1 BC to AD 22) some very clear statements were made on silt problems. The hydraulic engineer Zhang Rong, who flourished in 1 BC, was the first to give a numerical estimate of the silt content of loess-bearing water, and to point out that its rate of depositing a sediment was proportional to the speed of flow. His figures were sound, and inspired many later engineers.

When we look at modern studies, we are likely to feel some surprise at the degree of awareness of the old Chinese engineers, coupled with a sobering realisation of the prominent part still played by experimental observation and ideas three centuries after the Renaissance. Laws such as those which state that the line of deepest soundings in a river hugs the concave bank, opposite which sandbanks form, and that the depth increases with the amount of curvature, the Chinese already knew; they also were aware that, in general, silt deposit occurred more the slower the flow of water. Moreover, they consciously sought for conditions which would give a stable silt regime, so that a waterway would neither choke itself up nor erode in a way which would alter its course. They knew, too, that dykes were not as good as dredging, and that direction of flow should be arranged to encourage the natural scouring of sandbanks.

THE RIVER AND THE FORESTS

The greatest argument concerned the handling of the Yellow River on the North China Plain. The oldest attempts to control it in historical times seem to have been by the use of dykes built along the lower reaches under the superintendence of the Duke of Han. This was in the seventh century BC, and though it rests upon local tradition in the Han rather than upon documentary evidence, it is quite acceptable. His dykes had the effect of uniting the nine streams of the previous delta, and their remains still existed in Han times.

From the Qin and Han onwards, the greatest possible efforts were made to prevent the river from overrunning the plain during high discharge periods. After all, it must have been noticed that in the uncontrolled state the river tended to build up a kind of low bank on each side of its winter channel, and so for 2000 years the Yellow River had been enclosed with dykes, constantly increasing in dimensions (Fig. 467). Every few years the river would rise to levels which threatened to overflow them, or else the meanders of the low water channel would carry the river against an embankment and cause its collapse. Some fifty times or more the river has escaped from all control and formed a new channel on the plain, destroying in the process vast tracts of cultivation and settlement, and burying some land under excessive silt deposits. Thus, in the *Shi Ji* (Historical Records), which take us back to 99 BC, Sima Qian writes:

> The Han had been in power for thirty-nine years when, in the time of the emperor Xiao Wen Di, the River overflowed at Suanzao and broke through the 'Metal Dyke' (in 168 BC). . . .
>
> Rather more than forty years later, in the present reign, in the Yuanguan reign period (132 BC), the River again overflowed at Huzi, pouring off the south-east in the Zhuye marshes, and communicating with the Huai and Si rivers. The Son of Heaven therefore commissioned Ji Yan and Zheng Dangshi to recruit men to close the breach, but it suddenly opened again.

導黑水副圖

Fig. 467. A late Qing representation from *The* Historical Classic *with Illustrations* of AD 1905 showing river conservancy work. A dyke is being strengthened and sandbanks removed; baskets and ramming tools (punners) are seen in use.

The marquis of Wu-An, whose fiefs lay north of the river, urged that such great floods were direct acts of heaven with which it would be unwise to interfere. Nothing was done therefore for twenty years, but eventually the condition of the provinces became so bad that the emperor Wu Di made a tour of inspection himself. As a result, a breach at Xuanfang was successfully filled up in 109 BC, and a triumphal pavilion erected there. Yet it was a triumph destined to need perpetual renewal.

In spite of the centralisation of the Han bureaucracy, the level of planning and control for such projects still remained inadequate. Fifty years later, the great Han engineer, Jia Rang, who was a Daoist hydraulic engineer, believed that the great river should be given plenty of room to take whatever course it wanted. Rivers, he said, were like the mouths of infants – try to stop them up and they only yelled the louder or else were suffocated. 'Those who are good at controlling water give it the best opportunities to flow away, those who are good at controlling the people give them plenty of chance to talk.' He therefore recommended the wholesale resettlement of those living in regions bordering the river, and, if the emperor was unwilling to do this, he suggested that a great network of irrigation canals be made to relieve the flood pressure. This would require stouter sluice-gates than previously constructed. The only other alternative he could see was just to continue repairing dykes and embankments. At all events, there came an historic breakout of flooding and a change of the river's course in AD 11.

If there were engineers of Daoist tendency, there were also Confucian ones. Those who believed primarily in low dykes set far apart were opposed by those who believed in the main strength of high and mighty dykes, set nearer together. Equally, those who believed in giving a river's lower reaches the maximum degree of freedom were opposed by those who believed in contracting the channel so as to make the river dig its own bed. The former argued that with widely separated dykes there would be ample storage space between them for summer flow. The latter held that with a constricted channel, the water, flowing more rapidly, would itself scour out the sandbanks desired by the former group. The contractionists generally had the advantage of the expansionists because their plans, though more expensive, raised no difficult social problems of resettlement of population. But in so far as the control of the Yellow River today necessitated the construction and evacuation of vast retention basins alongside the main stream, the expansionists have not been entirely unjustified by modern technology.

However, for twenty centuries the two schools argued, and neither proved wholly successful. The deep channel is liable to approach and undermine the dykes at bends, and rises of water level occur with inconvenient speed. The Chinese were extremely conscious of such bend erosion, having a term (*kan* [*khan*] – a niche) for the scooped out bend. On the other hand, wide

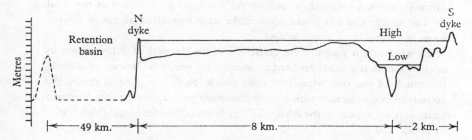

Fig. 468. Cross-section of the bed of the Yellow River just above the railway bridge linking modern Zhengzhou with Xinxiang, i.e. just west of the old Bian Canal.

separation of dykes allowed so much sediment to be deposited that storage capacity was reduced very quickly. So possession was taken of new land, and smaller parallel dykes were built to keep out ordinary floods, but this, of course, annulled the original objective.

At all events, it is generally agreed that the bed of the Yellow River has risen by nearly 1 metre per century, and in some places it has been necessary to build dykes up almost 2 metres per annum for stretches of some 30 to 50 kilometres to combat the huge silt deposits. Now the bed is only a metre or two below the level of the plain, and in some places it is either level with it, or even as much as 3.7 metres above it, but in cross-section we see that this central channel is very small compared with the kilometres of deposit on each side. Fig. 468 illustrates this. Consequently, when the river in high flood escapes through the dykes, the residual flow in the old channel fills that up with silt and plugs it against free flow in that course in the future. It is therefore very difficult to get it to resume its old course after a break. Neither of the ancient methods is really applicable, since dykes cannot be heightened indefinitely, and the work of removing a 7.5 metre or more mass of silt 1600 kilometres long and 8 kilometres wide is not feasible. The solution seems to be the retention basins, which are long strips of plain level 'wash land' parallel to the river with, on their outer sides, dykes massive and high enough to cope with any probable temporary influx of water.

The origin of these mountains of silt arises from the blanket of loess which extends for some 388 500 square kilometres over the drainage areas of the Yellow River, with an average depth of 30 metres, but varying from a few centimetres to 300 metres and more. As long as it was covered by forests of trees, bush and wild grasses, the soft soil was protected from the cutting effects of the heavy monsoon rains. But, with increasing population pressure, progressive deforestation, denudation and erosion has occurred. To-day, few trees are to be seen on the uplands of Shaanxi and Gansu. In north

Shaanxi, erosion gullies now occupy no less than 50 per cent of the total ground space, and the tracks often make their way along ridges of eroded loess as narrow as 1 metre across.

To what extent people were conscious of the dangers of deforestation in ancient China is a question hard to answer, for one must beware of reading too much of our own ideas into their words. Still, it is hard to doubt the conservation significance which has traditionally been ascribed to the Mencian adages which appear in the *Meng Zi* (The Book of Master Meng (Mencius)) of about 120 BC:

> If the seasons of husbandry be not interfered with, the grain will be more than can be eaten. If close-mesh nets are prohibited in the pools and lakes, the fishes and turtles will be more than can be consumed. If axes and hatchets are used in the mountain forests only at suitable times, there will be more wood than people know what to do with.

Elsewhere Mencius compares deforestation to the forcible debauchery of the natural goodness inherent in the human race. Indeed, it is clear that he regarded the denuded state as artificial and ominous, and he recognised also the danger of overgrazing.

All the same, the destruction of forests must have been severe in ancient times because they burned clearings before planting. To this the *Huai Nan Zi* (The Book (of the Prince) of Huainan) of 120 BC, in one of its diatribes against the luxurious decadence of feudalism as opposed to the communalism practised in the golden age, adds deforestation for the needs of fuel for metallurgy as yet another evil.

By the sixteenth century AD, the direct relationship between denudation, erosion and flood problems was well recognised. The *Shanxi Tong Zhi* (Provincial Historical Geography of Shanxi), edited by Luo Shilin in AD 1733, quotes the Ming scholar, Yan Shenfang:

> Before the Zhengde reign period [AD 1506 to 1521] flourishing woods covered the southeastern slopes of the Shangchi and Xiazhi mountains. They were not stripped because the people gathered little fuel. Springs flowed . . . entered the Fen River at Shangduanduo as the Zhangyuan River. . . . It was never seen dry at any time of the year.
>
> But at the beginning of the Jiajing reign period [AD 1522 to 1566] people vied with each other in building houses, and wood was cut from the southern mountains [in the Qi district of Shanxi] without a year's rest. People took advantage of the barren mountain surface and converted it into farms. Small bushes and

seedlings in every square metre of ground were uprooted. The result was that if the heavens send down torrential rain, there is nothing to obstruct the flow of the water. In the morning it falls on the southern mountains; in the evening when it reaches the plains, its angry waves swell in volume and break through the embankments, changing the course of the river. . . . Hence the district of Qi was deprived of its wealth.

Even so, the balance swung between frugality and extravagant use of natural resources during the successive centuries, though Chinese farmers, at any rate, had always been very conscious of the need for skillful use of soil moisture. Terracing and other methods that retained water in upland soils, and so helped the water conservancy problem of the lower river valleys, were indispensable.

The climate of northern China consists of a dry and windy spring, a hot dry summer with showers at long intervals, a very wet late autumn bringing two-thirds of the whole year's rainfall and causing great erosion if not serious floods, then finally a severe dry winter with wind-borne snow. Farming in these conditions could only be successful if everything possible was done to conserve water in the soil. And, indeed, the oldest agricultural books such as the *Fan Sheng Zhi Shu* (The Book of Fan Shengzhi on Agriculture), written about 10 BC, and the *Qi Min Yao Shu* (Important Arts for the People's Welfare), of about AD 540, are full of recommendations about means of collecting and conserving soil moisture. The right times for ploughing, hoeing and harrowing, the ways of retaining snow and even dew, 'dust mulch' (a fine loose layer of surface soil maintained by cultivation) and leafage canopy are all intelligible in terms of reducing evaporation and retaining water.

One of the oldest systems for doing this on sloping ground was pit cultivation, that is the cultivation of crops in shallow pits and ditches. As The Book of Fan Shengzhi on Agriculture points out, this depends mainly on the fertilising power of the soil in the pits, so good land is not at all necessary; mountain sides, the edges of cliffs, steep places near villages, and even the inside slopes of city ramparts, can all be used. Pit construction can start directly on waste land, and the book goes on to give the technique in elaborate detail. Clearly, this method partook of the nature of very small terraces and must have contributed greatly to the reduction of rapid run-off.

CIVIL ENGINEERING AND ITS SOCIAL ASPECTS IN THE CORPUS OF LEGEND

Yu the Great controlled the waters – such is the phrase which has become proverbial, and which one encounters all over China. Probably no other people in the world have preserved a mass of legendary material into which it is so clearly possible to trace back engineering problems of remote times. The essential facts of the story are that two culture heroes of civil engineering were successively put in charge of regulating and controlling the floods and rivers by the legendary emperors; the first, Gun, failed, but the second, Yu, succeeded. Many of the incidental features of this story which have come down to us are so revealing that they are well worth examining.

In the time of the legendary emperor Yao, there were great floods, and Gun was put in charge of all works of protection and control. But after nine years his efforts were unsuccessful; the water rose as fast as he could build his dykes. His efforts were considered a complete failure and he was banished, killed by Yao, and his body was cut into pieces. Later, Gun was regarded as the patron and inventor of dykes, embankments and walls.

Afterwards, Yu, the son of Gun, was appointed by Shun (Yao's successor), and within thirteen years, by titanic labours, he opened the courses of the nine rivers, conducting them to the four seas and deepening the canals. The motive of the dredging of waterways is always connected with Yu, who was supposed to have been fully successful in his work. He came to occupy the permanent position of irrigation culture-hero which Gun had not attained. Besides these two chief characters, there was a subsidiary figure, namely Gonggong (literally 'communal labour') whose other name was Chui. Recommended to Yao as Controller of Waters, he was later rejected by Shun, and was finally banished and killed.

There is method in all this, if rightly interpreted. It has been suggested that the contrast between Gun and Yu betrays the existence of two rival schools of hydraulic engineering thought in ancient China. This was part of the conflict which has existed throughout Chinese history between partisans of high dykes and those in favour of deep channels. Moreover, it took the form of a conflict between two systems of morality, one in favour of confining and repressing Nature, the other in letting Nature take her course, or even assisting her to return to it if necessary.

Not every school of engineering thought has had its maxims engraved on stone and venerated for centuries. Yet such is the case for the followers of Yu the Great. At the town of Guanxian, some distance north-west of Chengdu, the capital of Sichuan, there exists what may be one of the most remarkable irrigation works in the world. A great river, the Min Jiang, comes tumbling out of the mountainous country of the Tibetan borderland, and is diverted at certain times of the year by movable dams and spillways into an enormous cutting through the hillside, forming thereby an artificial river which fertilises over 2000 square kilometres of first class agricultural land by means of almost 1183 kilometres of irrigation channels. The point here is that it dates from the middle of the third century BC, having been started by

Li Bing, then the governor of the province, and completed by his son, Li Erlang. In the temple dedicated to the latter, there are a series of stone-cut inscriptions, probably of no great age themselves, but perpetuating certain old key phrases, some attributable to the Qin, such as 'Dig the channel deep, and keep the dykes (and spillways) low', and 'Where the channel runs straight, dredge the centre of it; where it curves, cut off the corners'.

Relations with afforestation in the legends are also curious. Ancient ideas of vegetation covered hills have already been mentioned (see p. 188). But the traditions make Yu the 'attacker' of forests, and conceivably forest removal may have been to reduce rainfall and so ease the tasks of the engineers lower down, for when evaporation on the windward side is the chief cause of rain, deforestation may, to some extent, reduce rainfall. Yet the fearful erosion produced far outweighs any slight reduction of rain it could cause, and there seems to be some conflict of opinion. For instance, when in 525 BC envoys were sent to make sacrifice in time of drought, many trees were cut down, but the rationalist statesman Gongsun Qiao said this was a great evil and had the perpetrators punished. Yet later, in 219 BC, the first emperor Qin Shi Huang Di, who had conceived a grudge against Mount Xiang, had it completely deforested.

THE FORMATIVE PHASES OF ENGINEERING ART

There is a classification of irrigation canals in the *Record of the Institutions of the Zhou (Dynasty)* (*c.* 1030 to 221 BC); this gives their dimensions, though the literary tradition seems rather stylised, for the figures are doubled at each step, thus producing cross-sections which were then impracticably deep. One thing it does raise, however, is the 'well-field' system, which really signifies in China a pattern of land distribution, and more properly belongs to questions of land tenure; nevertheless it has some connection with the origins of irrigation.

In the sixth century BC, though Confucius (551 to 479 BC) makes no mention of large irrigation works, he does refer to 'ditches and conduits', yet it was not until about 300 BC that Mencius gave the classical description of the well-field (*jing tian* [*ching thien*]) system. However, water supply is never far from the meanings of the words used, for the word *jing* (井) has two distinct meanings – its usual one of a well and the technical one of a nine-plot land division. It is therefore intimately connected with *tian*, the fields (田). Yet some ninth century BC bronze forms of the character for *jing* actually indicate a well or water-source by a dot in the centre, and such dots also appear within the spaces of *tian*.

All this makes it easier to understand what the writer of the *Record of the Institutions of the Zhou (Dynasty)* was talking about when the author built up a series of technical irrigation terms from the nine-plot *jing* or well-field. By this time, the well itself had long evaporated to a mere symbol, and the fields were supplied with water according to a system more central or southern than northern in character, even though Han legend attributed even the refinements of this to the teaching of Yu the Great himself.

But whatever apportionment of land there is, the fact remains that the basic social requirement for large scale water control was the possibility of extending the ancient *corvée* system, whereby work was done on the lord's field, to a vast mobilisation of labour for public works. So long as peasants were attached to well-field units, there could be no great number of surplus and unattached labourers. But the system was already breaking down at the beginning of the sixth century BC, for in 539 BC the State of Lu began to levy a tax according to the area occupied, irrespective of the identity of the holder. Such a tax would supersede a direct contribution of labour on the lord's land, and, as productivity of labour increased, there gradually grew up a large labouring population which, at any rate at certain times of the year, could be withdrawn from agriculture for the construction of great engineering works.

In other words, it was the coming of private land ownership, together with an increase of agricultural productivity in the New Iron Age, which 'liberated' the labour force required for the greater works. The first State to recognise the decay of the well-field system and to hasten its disappearance was the Qin, which abolished it in 350 BC. The transfer of land by sale and purchase was legalised, especially for the benefit of commoners who had performed meritorious military services. Lots of any size might be held by smaller or larger land-owning families. Then, in 348 BC, a land tax was established. Thus began the perennial pattern of feudal bureaucracy, which all other regions were to follow. It can hardly be a coincidence that this change was accompanied by exceptional development of successful irrigation and transportation projects (shortly to be described).

One cannot overemphasise the role of manpower in the ancient and medieval achievements of Chinese civil engineering. Work in those 'eotechnic' ages had to be done somewhat on the basis of 'a million men with teaspoons'. This is illustrated in traditional style in Fig. 469, which is taken from the *Hong Xue Yin Yuan Tu Ji* (Illustrated Records of Memoirs of the Events which had to happen in my Life) (AD 1849), by Linqing, himself an able director of engineering works.

The geo-climatic conditions of China exerted an irresistible influence on Chinese society in the direction of strengthening centralised government. The simple reason for this was that any effective treatment of the engineering problems set by the rivers, and the desired intercommunication between watercourses, tended at every stage to cross the boundaries of the smaller feudal units. But crossing boundaries was not the only factor. At an early stage in the Warring States period, it was realised that water might be a weapon. The sight of rooftops swirling round in flood waters, with people clinging to them amidst floating trees and bodies of dead animals, appears likely to have suggested to the feudal States that a strategic construction or destruction of dykes and watercourses might bring about gratifying results. Moreover, States competed violently with one another for irrigation water, and in appropriating drained areas for growing crops.

It is not surprising, then, that all subsequent governments controlled rivers and canals as parts of a unified system, their success in achieving this varying both with political power, the extent of the realm and the arbitrary conditions set by natural surroundings. But in times of war, both civil and foreign, there were often temptations to undo what had so laboriously been done. For instance, in AD 923, the Later Liang general in the field against the Later Tang broke Yellow River dykes deliberately, and the flood weapon was again used in 1128. All this, of course, is to say nothing of the motivation to move military supplies by the construction of great transport canals.

SKETCH OF A GENERAL HISTORY OF OPERATIONS

One of the songs in the *Book of Odes* contains the words 'How the water from the Biao pool flows away to the north! Flooding the rice fields . . .'. Since this may indeed date from the eighth century BC, it is perhaps one of the earliest mentions of irrigation in Chinese history, but it can only refer to a reservoir on some rather small scale. Yet a couple of centuries later, impressive works were well underway.

One of the larger, or at least longer, works of Chinese hydraulic engineering, however, had the reputation of being so ancient that its first origins were lost in the depths of time. This was the Hong Gou – the Canal of the Wild Geese, or Far-Flung Conduit – which connected the Yellow River near Kaifeng with the Bian and Si rivers, and ultimately formed a model for part of the Sui Grand Canal.

Though the conduit may have been first made to bring irrigation water to the upper Huai basin rather than for transport, its linking of the Huai and Yellow Rivers at an early date allowed barges to travel the east-central and northern key economic areas. The watershed between the Yellow River and the Huai basins is very low and flat, and what difference there was between levels was lessened to almost nothing by the building up of the bed of the Yellow River, already advanced in Sima Qian's time (about 90 BC).

Strictly speaking, this Canal of the Wild Geese was a complex of artificial waterways rather than a single canal. It took off eastwards from the Yellow River at a point downstream from the entry of the Luo River from Luoyang, just beside the city of Rongyang, where there was a great granary in Han times and which was the hub of the tax-grain transport system. It then

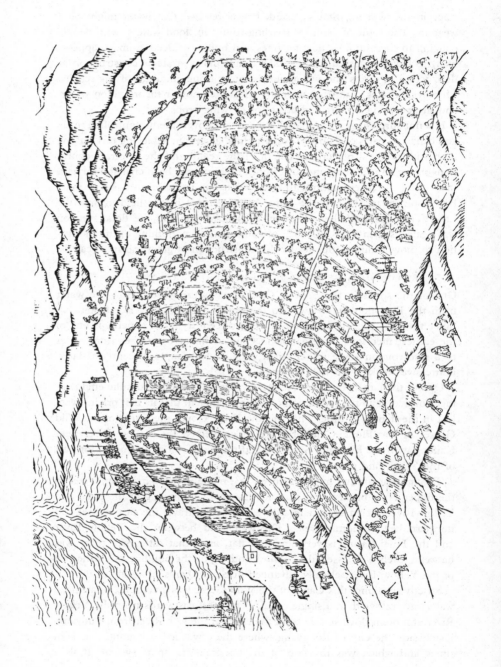

swung south in an arc of almost 420 kilometres so as to connect with the parallel south-eastward flowing rivers entering the Huai from the north – the He, the Zi, the Sui, the Guo and the Ying. This canal, called the Langdang Qu, joined together the upper reaches of the last three of these rivers, though it was never the part most used for traffic. That function devolved on its most northerly branch, the Bian Canal or River, some 800 kilometres long, which joined the head-waters of the He, and so, passing near Xuzhou, gave access to the Huai.

From AD 70 onwards the canal's mouth on the Yellow River was protected by substantial dykes built under the supervision of Wang Jing (*died* AD 83). This then was the Canal of the Wild Geese proper, or the Bian Canal of the centuries before AD 600. Yet the question of its construction is still very uncertain. So old was it supposed to be that Sima Qian gave it pride of place after the works of the Great Yu himself, assuming no clear origin to it. The diplomat Su Qin mentioned it in about 330 BC, and some historians believe it to have been built between 361 and 353 BC, yet if so it is curious that Sima Qian did not know about its connection with Prince Hui of Wei. Alternatively, the sixth century BC or the early part of the fifth century – the time of Confucius himself – might be a reasonable estimate,

---

Fig. 469. The Chinese genius for organising very large numbers of workers in civil engineering operations is illustrated by a unique drawing from the autobiography of the high official and hydraulic engineer Linqing, his *Illustrated Records of Memoirs of the Events which had to happen in my Life*, of AD 1849. The drawing is probably by his friend Chen Jian. The somewhat enigmatic caption ' "Wrestling for the Red" when Cutting a Canal' is explained in an accompanying text. When the job is more than half done, the superintendents begin to 'hang up the red', i.e. to organise competitions of speed of working for prizes of meat and wine, of boots and hats. When it is nine-tenths done they all set up a large umbrella-like lantern of red silk as a thank-offering to the local gods for the absence of mishaps, and upon the lantern the names of the prizewinners are inscribed. This was an old custom, Linqing said, but far better than the threats and favouritism of some officials, which led only to strikes. The actual work shown in progress was the cutting of a canal at Zhongmou, a place south of the Yellow River between Kaifeng and Zhengzhou in Henan, doubtless between AD 1833 and 1842 while Linqing was Director-General of River Conservancy. Apart from many wheelbarrows, one can see more than a dozen square-pallet chain-pumps (see p. 263), these being apparently hand-worked. Other men are using hand-swung buckets for draining and excavation. At the top on the left are a 'theodolite', levelling staff and chain-measure. Further left, the engineer-in-charge, on a visit of inspection, is riding a horse on the neck of land which will be washed away when the canal is opened. In the foreground there is an altar to the local deity, with some guards on one side and a group of old people on the other.

for Sima Qian does not mention the State of Wei, which was not founded until 403 BC in connection with the canal, but only smaller and more ancient States. It is clear, then, that a Confucian dating is by no means excluded.

The date of the earliest known irrigation river, though prior to this, can be accepted with little reserve. In northern Anhui, south of the city of Shouxian, there still exists a great tank, almost 100 kilometres in circumference, and anciently known as the Peony Dam (Shao Bei). Both Sima Qian and the *The Book (of the Prince of) Huainan* of 120 BC make it clear that this had been built under the superintendence of the Sunshu Ao, minister to the Duke Zhuang of the Chu State between 608 and 586 BC. The dam simply flooded a rather flat valley of no less than 24 000 square kilometres with water from the Yangtze, and was repeatedly repaired during the Han and Tang dynasties.

In the fifth century BC there were a number of hydraulic works; two were of special importance. First, Ximen Bao, the rationalist statesman who abolished human sacrifices to river gods, organised a diversion of the Zhang River, leading it away to irrigate a large region of Hebei instead of adding wastefully to the burden of the Yellow River, though it was not completed until 318 BC or soon thereafter. The second was the Han Gou (Han Country Conduit), which connected the Han River with the Yangtze, which when later altered became the second oldest segment of the Grand Canal.

During the Qin, there was a famous system in Sichuan built about 270 BC and also the Zhengguo Canal (about 246 BC), to be described later. A third Qin project, less well known, was near Ningxia, where the Yellow River traverses a broad valley and the land on both sides is fertile if irrigated. It was a splendid outpost against the Huns, and in 215 BC Qin Shi Huang Di sent his general Meng Tian to take possession of the area, and the next year lateral irrigation canals were built; these now cover an area of about 160 kilometres long and 30 kilometres broad.

Coming to the Han period, there was much civil engineering activity during the time of Han Wu Di. By then the grain tribute coming from the east of the passes had enormously increased, and in 133 BC the hydraulic engineer Zheng Dangshi proposed a new canal of about 160 kilometres in length. He said:

> If we take water from the Wei River to fill a canal dug from
> Chang'an to the Yellow River, running along the foot of the
> southern mountains for a little more than 300 *li* (in a straight line),
> transportation will be greatly eased. Besides, the people living
> below the canal would be able to irrigate more than 10 000 *qing*
> [*chhing*] (about 672 square kilometres). Thus in one sweep we
> shall shorten the time taken by the (grain) transports, reduce the

number of men required, augment the fertility of the lands within the passes, and obtain fine harvests.

The emperor approved the project, and it was completed within three years. Seven centuries later, the Sui engineers reconstructed this canal and used it as part of their great system, whereby Chang'an was in direct water communication with Hangzhou, though we do not know whether they used exactly the same course. In any case, the Sui's Cao Qu (Grain Traffic Waterway, *par excellence*) remains the classical Chinese example of the lateral transport canal.

Not all the works of the Early Han period were equally successful, however, but towards the end of the dynasty, between 38 and 34 BC, a fine work was carried out in southern Henan by Shao Xinchen, the prefect of Lingling and Nanyang. This was the Qianlu Bei (reservoir), which dammed up one of the larger northern tributaries of the Han River and irrigated 30 000 *qing* (over 2000 square kilometres). It is of interest for several reasons. First, Shao introduced six sluice-gates set in stone which greatly aided distribution of the water. He also introduced equitable distribution of water for the farmers by appointing mutual restraints, and indeed he became so popular that he was known as 'Father Shao'. Secondly, one of his successors in office, Du Shi, equally beloved, was the man who in AD 31 introduced the use of waterwheels for metallurgical blowing-engines; hence, the name of the dam, the 'Tongs-and-Furnaces Dam', can hardly be a coincidence. Nanyang remained for a long time the centre of an iron industry which employed water-power.

Though the later Han dynasty was less active in hydraulic projects, valuable things were done. The road-builder Wang Jing constructed the Peony Dam between AD 78 and 83, but his greatest work, carried out with Wang Wu in the previous decade, was the thorough reconstruction of the Bian Canal, with the introduction of numerous flash-lock gates. A century later, Chen Deng built a series of weirs from the city of Shouxian westwards, collecting the water from thirty-six streams over an area of some 30 kilometres in diameter, and irrigating 10 000 *qing* (400 square kilometres). The remains of these structures came to light in 1959, and it appears that they were made of alternate layers of rice straw and clayey soil based on a gravel foundation, the stalks being laid parallel to the current flow, and the whole supported by wood piling and coffering, denser at the centre than at the ends. Military canals were also built at this time.

The Grand Canal as a unit was the creation first of the Sui (AD 581 to 618), when it led to Luoyang, and then of the Yuan (AD 1271 to 1368), when it led to Beijing. But before treating it in detail (p. 218), since space is limited, it will be useful to summarise the work done from Zhou and Qin, that is from about the seventh to third centuries BC up to the Qing (or Manchu) dynasty, which came to its end in AD 1911. This may best be done briefly using Fig. 470, which, it must be noted, uses a vertical logarithmic scale. It indicates that for the earliest periods, where records are sparse, there is little doubt about a start being made, though a real concentration on hydraulic engineering did not commence until the Former Han. The Three Kingdoms period (AD 220 to 280) shows a distinct volume of activity, doubtless for strategic reasons. Continued unsettlement from the fourth to the seventh centuries AD led to few works by the smaller dynasties of the time, but there was a leap during the Sui, especially with regard to the Grand Canal, and this was sustained during the Tang and the Song. It is also evident that there was more activity after the removal of the capital to the south in the twelfth century AD than before, because nomadic invasions seem to have stimulated more intensive cultivation south of the Yangtze, though the 'nomadic' empires never understood the vital importance of hydraulic engineering as much as the Chinese dynasties did. Heightened technical ability from the fifteenth century AD onwards is seen in the high figures for the Ming and Qing. With this, then, we may begin a tour of inspection of the outstanding projects of the past.

THE GREATER WORKS

Among the most notable works of Chinese civil engineering, there are three of Qin date (221 to 207 BC). They are the Zhengguo Irrigation Canal, the Guanxian Irrigation System and the Ling Qu or 'Magic Transport Canal'. The first two were part of the organisation of the key economic areas (north-western and western), and the third was a brilliant achievement connecting the north and centre with the extreme southern regions. These compare in greatness of conception, if not length, with the Grand Canal itself – a work mainly of Sui and Yuan times (the seventh and thirteenth centuries AD).

The Zhengguo Irrigation Canal (Qin)

The story of this canal is carefully told both in the *Historical Records* of about 90 BC by Sima Qian and his father Sima Tan, and in the *History of the Former Han Dynasty*, which appeared about AD 100.

> (The Prince of) Han, hearing that the State of Qin was eager to
> adventure profitable enterprises, desired to exhaust it (with heavy
> activities), so that it should not start expanding to the east (and
> making attacks on the Han). He therefore sent the hydraulic
> engineer Zheng Guo (to Qin) to persuade deceitfully (the King of)
> Qin to open a canal from the Jing River, from Zhongshan and
> Hukou in the west, all along the foot of the northern mountains,

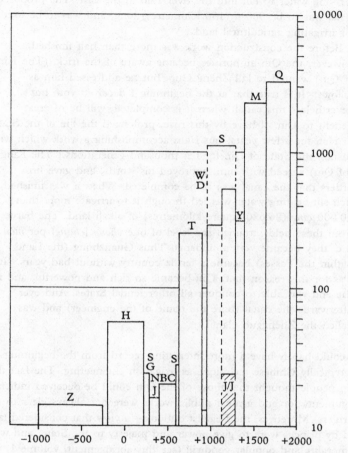

Fig. 470. Plot of the number of hydraulic engineering works undertaken in successive dynasties: Z = Zhou, H = Han; S =San Guo; J = Jin; NBC = Nan Bei Chao; S = Sui; T = Tang; WD = Wu Dai; S = Song; J/J = Jurchen Jin; Y = Yuan; M = Ming; Q = Qing. The horizontal scale is chronological, where a minus sign denotes dates BC, and a plus sign denotes dates AD. The total Song figure is greater than its two component columns for the Northern and Southern Song periods, respectively, because there were a certain number of projects which could not positively be allotted to either.

carrying water to fall into the river Luo in the east. The proposed canal was to be 300 *li* (150 kilometres) long, and was to be used for irrigating agricultural land.

Before the construction work was more than half finished, however, the Qin authorities became aware of the trick. (The King of Qin) wanted to kill Zheng Guo, but he addressed him as follows: 'It is true that at the beginning I deceived you, but nevertheless this canal, when it is completed, will be of great benefit to Qin. [I have by this ruse, prolonged the life of the State of Han for a few years, but I am accomplishing a work which will sustain the State of Qin for ten thousand generations.] The King (of Qin) agreed with him, approved his words, and gave firm orders that the canal was to be completed. When it was finished, rich silt-bearing water was led through it to irrigate more than 40 000 *qing* (26 680 square kilometres) of alkali land. The harvests from these fields attained the level of one *zhong* [*chung*] per *mou* (i.e. they became very abundant). Thus Guanzhong (the Land within the Passes) became a fertile country without bad years. (It was for this reason that) Qin became so rich and powerful, and in the end was able to conquer all other feudal States. And ever afterwards the canal (bore the name of the engineer) and was called the Zhengguo Canal.

One could hardly have a more interesting record from the beginnings of such a typically Chinese technique as irrigation engineering. The fact that the Han people thought that those of the Qin could be deceived indicates that arguments for and against public works were not always clear to the feudal rulers. Moreover, the Legalist outlook – a view that considered laws enacted by princes were to give power and majesty to the State, and were above morality and popular goodwill (see this abridgement, volume 1, pp. 273 ff.) – was at its height at this time. The willingness of the Qin State to accept this approach led it to its subsequent growth in power. Yet it was perhaps this attitude which suggested to the Han that the Qin ruler would easily swallow a new and grandiose irrigation project even though it might well appear of doubtful success. But there is no reason to think that Zheng Guo sabotaged it; he seems to have been professionally honest, and perhaps only realised himself during the course of the work what it would mean to the Qin once successfully completed. Sima Qian understood perfectly the fundamental importance of increased productivity of food, as well as the supply potential of such a scheme for the ultimate political success of the Qin. Indeed, he must be regarded as one of the first historians to appreciate that the supply potential is at least as important as military power in great

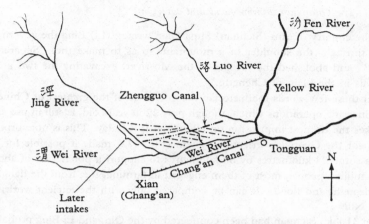

Fig. 471. Sketch-map of the Zhengguo Canal irrigation system, first completed in 246 BC and still now in use under the name Wei Bei, or Jinghui Qu, Irrigation Area. The small south-flowing river within the area is the Sichuan He. The Luo River now falls to the west of the Wei, and not directly into the Yellow River.

conflicts. What is more, he was fairly close in time to the events in question, for he laid down his brush in 90 BC, and Zheng Guo's work had been brought to completion only some 150 years before, in 246 BC, the year which witnessed the crowning of the future First Emperor as King of Qin.

Only a few years before Sima Qian was writing, extensive additions were made to the Zhengguo Canal in 111 BC, supplementary lateral contour branches being used to irrigate land higher than the main canal. Then, in 95 BC, the high official Bai Gong pointed out that the canal had become so silted up as to lose much of its value, and he proposed to tap the Jing River a good deal higher up, and to carry a new canal for some 96 kilometres following a contour above that of its predecessor. This was successfully accomplished.

Recutting the canals and tapping the Jing River higher up was an ongoing process for the next twenty centuries. This whole Wei Bei irrigation area (shown in Fig. 471) is unusual in that, although it was one of the first really large projects in China, it has existed in use to the present day. The constant recutting is due to the constant battle with silt, and the reason for moving the intake higher and higher up the Jing is that the river has been continuously eroding its bed. Over the years, various modifications have been made, and today the intake is far up the Jing gorges where the river has a rocky bed. The original Qin or Han intake is still identifiable, but this point is no less than 15 metres above the present level of the river.

The Guanxian division-head and cut (Qin)

In the country of Shu (Sichuan) Sima Qian wrote: '(Li) Bing the governor, cut through (the shoulder of a mountain, so as to make the) "Separated Hill", and abolished the ravages of the Mo river, excavating the two great canals in the plain of Chengdu.'

In these few words the historian recorded one of the greatest of Chinese engineering operations which, though now 2200 years old, is still in use and makes the deepest impression on all who visit it today. This is not surprising, for the Guanxian irrigation system (Fig. 472) made it possible for an area some 64 kilometres by 80 kilometres to support a population of about five million people, most of them engaged in farming, free from the dangers of drought and floods. It can be compared only with the ancient works of the Nile.

In 316 BC, Sichuan had been conquered by the Qin, and Li Bing probably helped to fortify its cities in 309 BC. In 250 BC, Prince Xiao Wen appointed him governor of the province, but as it is not likely that he lived long after about 240 BC, the great works at Guanxian were probably completed under the supervision of his son Li Erlang, in about 230 BC, and there can be no doubt that the project they carried through was, like the Zhengguo Canal, one of the great sources of the strength of the Qin State and Empire.

At Guanxian, the Min Jiang River flows into the basin of Sichuan from its source in the hills of the extreme north of the province surrounding Songpan. Li Bing decided to divide it into two great Feeder Canals, the Inner and the Outer, by means of a division-head of piled stones, known as the Fish Snout (Fig. 473), about the point where the famous suspension bridge (see Figs. 458 and 460) crosses the river. The drawing in Fig. 472 gives the general plan, and the model in Fig. 474 depicts a panoramic view, while Fig. 475 shows a general view of the bridge and the head-works. In recent times, by 1958, so many new distribution canals had been built that some 3764 square kilometres were being supplied with water, and the Authority's estimate was that, when all the possible extensions of Li Bing's system were complete, no less than 17 800 square kilometres would be irrigated.

The works, dams and spillways are known as the Dujiang Yan, or the Dam on the Capital's River. Its primary feature is the division-head of piled stones which separates the two feeder canals. The inner one is used wholly for irrigation, while the outer, which follows the old course of the river, acts as a flood channel as well, and also carries some boat traffic. In order to construct the inner canal, Li Bing had to make a vast cut through rock at the end of a range of hills, on part of which the city of Guanxian is built. This is known as the 'Cornucopia Channel', and the height of the 'Separated

Hill' on which the temple dedicated to him now stands (Fig. 476) is some 27 metres from the canal bed, so that the total height of the 27 metre-wide cut would have been almost 40 metres (Fig. 477). The primary division-head and the rock cutting the two channels are the Diamond Dyke (Fig. 472*f*, *f*) and the Flying Sands Spillway (Fig. 472*h*). The top of the Diamond Dyke is made higher than the flood level of the Min River in order to assist division of the water, that of the Flying Sands Spillway is adjusted to the elevation required for the optimum supply of irrigation water to the inner canal. When floodwaters rise above this level, they overflow and automatically regulate the flow in the Inner Feeder Canal.

Immediately after passing the city of Guanxian, the inner canal begins to give off its side channels and sub-channels, of which there are in all 526 and 2200, respectively. Some of them pass through and beside the city of Chengdu, and all ultimately find their way with the Min River into the Yangtze at a point past Jiading.

Each year there is a cycle of operations corresponding to the flow of water. From December to March the river is at the low water stage, with an average flow of 200 cumecs (cubic metres per second), falling sometimes to 130. From April onwards, when planting starts, the flow increases, till at 585 cumecs the full requirements of the works of both outer canal (280 cumecs) and the inner one (305 cumecs) are satisfied. In summer, i.e. June and July, high water is reached, with a maximum flow of 7500 cumecs in the river as a whole, after which there is a slow decline until November, and then faster thereafter.

Throughout the centuries, the advice of Li Bing to clear out the beds and keep the dykes and spillways low has been faithfully followed, and if it has been possible to preserve the system to remain almost as he left it, this is partly because the river is not extremely silt laden, and partly because its annual fluctuations have allowed incessant and effective maintenance. Every year, about mid-October, the annual repairs begin. A long row of weighted wooden tripods is placed across the inlet of the Outer Feeder Canal and covered with bamboo matting plastered with mud to form a cofferdam, thus diverting all the flow into the inner canal. The bed of the outer canal is then excavated very actively to a predetermined depth, and any necessary repairs are made to the division-heads. About mid-February, the cofferdam is removed and re-erected at the intake of the inner canal, and similar maintenance of the inner system is carried out. On the fifth of April the ceremonial removal of the cofferdam marks the opening of the irrigation system and gives an excuse for celebration.

The only technical detail not yet mentioned is that, at various times, 'nilometers' or gauges for water-levels were incorporated at suitable positions. One of these was a human figure inscribed 'In the dry season let the

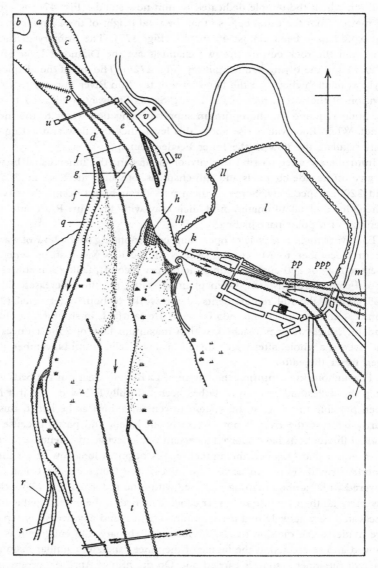

Fig. 472. Plan of the Guanxian irrigation system headworks, drawn by
Dujiang Yan. The intake works shown here beside the city and temples
of Guanxian distribute water in a myriad of canals all over the plain
of Chengdu in Sichuan.

Key to Fig. 472:

a, a	Min Jiang[1] (Min River)	*o*	Zouma He[18] (derivatory canal). New sluice-gates were installed across this, as shown, in 1952
b	Hanjia Ju[2] (Han-family island)		
c	Baizhang Ti[3] (Thousand-foot Dyke)		
d	Wai Jiang[4] (Outer Feeder Canal; old course of river)	*p*	Dujiang Yu Zui[19] (Fish Snout; primary divison-head of piled stones)
e	Nei Jiang[5] (Inner Feeder Canal)	*pp*	Taiping Yu Zui[20] (left secondary division-head)
f, f	Jingang Ti[6] (Diamond Dyke)		
g	Pingshui Cao[7] (Water-level Bypass or Adjusting Flume)	*ppp*	Ding Gong Yu Zui[21] (left tertiary division-head)
h	Feisha Yan[8] (Flying Sands Spillway)	*q*	Shahei Zong He[22] (right main derivatory canal)
i	Renzi Ti[9] (V-shaped Spillway)	*r*	Shagou He[23] (derivatory canal)
j	Lidui[10] (Separated Hill) and Fu Long Guan[11] (Tamed-Dragon Temple; the votive temple of Li Bing)	*s*	Heishi He[24] (derivatory canal). The feed into both of these is assisted by a spillway higher up, the overflow rejoining the Zhengnan Jiang
k	Baoping Kou[12] (Cornucopia Channel; the rock cut)		
l	Guanxian[13] City	*t*	Zhengnan Jiang[25] (old course of river, flood course, etc.)
ll	Yulei Shan[14] (Jade Rampart Mountain)	*u*	Anlan Suo Qiao[26] (suspension bridge, see Fig. 458)
lll	Fenglou Wo[15] (Phoenix Nest Cliff)	*v*	Er Wang Miao[27] (Temple of the Second Prince; the votive temple of Li Erlang)
m	Puyang He[16] (derivatory canal)		
n	Botiao He[17] (derivatory canal)	*w*	Yu Wang Gong[28] (Temple of Yu the Great)

Note (i). The two lines, drawn in the convention usual for railways, which connect the Thousand-foot Dyke (*c*) and the right bank of the Min R., respectively, with the Fish Snout primary division-head (*p*), represent the positions of the temporary *ma cha* dams set up when the water is low for clearing the beds of the Nei Jiang (*e*) and the Wai Jiang, (*d*), the former in January, the latter in November.

Note (ii). A steel cable, anchored at the point marked with an asterisk, guides rafts, floating tree-trunks, etc., into the Puyang He (*m*), avoiding the Zouma He (*o*).

Note (iii). On the hill above the Temple of the Second Prince (*v*), there is a small but beautiful Daoist temple to Lao Zi. An inscription says: 'The highest excellence does not lie in the highest place; In changes and transformations let nothing be contrary to Nature.'

[1] 岷江	[2] 韓家埧	[3] 百丈堤	[4] 外江	[5] 內江
[6] 金剛堤	[7] 平水槽	[8] 飛沙堰	[9] 人字堤	[10] 離堆
[11] 伏龍觀	[12] 寶瓶口	[13] 灌縣	[14] 玉壘山	[15] 鳳樓窩
[16] 蒲陽河	[17] 柏條河	[18] 走馬河	[19] 都江魚嘴	[20] 太平魚嘴
[21] 丁公魚嘴	[22] 沙黑總河	[23] 沙溝河	[24] 黑石河	
[25] 正南江	[26] 安瀾索橋	[27] 二王廟	[28] 禹王宮	

Fig. 473. The main divison-head (Dujiang Yu Zui or Yu Zui) at Guanxian, seen from the point where the suspension bridge crosses the artificial peninsula or island of the Diamond Dyke. Across the river lies the Thousand-foot Dyke. (Photo. Joseph Needham, 1958.)

feet be covered, in flood let the level not pass the waist'. This referred to the proper height for the slipways. The ancient iron bars of the Inner Channel were there for the same purpose, and at various points there are old measuring scales from which the water-levels in the various canals can be observed daily.

These gauges referred to, among other things, a stone inscription in the temple of Li Erlang – still to be seen today. Entitled *Zhi He San Zi Jing* (The Trimetrical Classic of River Control) it makes reference to 'The Six-Character Teaching' and the 'Eight-Character Rule' and those other Guanxian inscriptions previously mentioned (p. 191), but the date of these is unknown, though it will hardly be earlier than the thirteenth century AD. It was then that Wang Yinglin produced his *San Zi Jing* (Trimetrical Primer), a classical Confucian primer for schoolboys, in rhyming couplets for easy memorisation, and used down to modern times. The inscription runs:

'Dig the channel deep,
And keep the spillways low';
This Six-Character Teaching
Holds good for a thousand autumns.

Fig. 474. Model of the Guanxian head-works in the exhibition room of the Authority at the back of Li Bing's temple on the Lidui Hill. The Min River can be seen dividing near the suspension bridge into the Nei Jiang on the right and the Wai Jiang on the left. The Cornucopia Channel cutting is well shown between the old city walls to the right and the temple of Li Bing to the left. Li Erlang's temple appears to the right of the suspension bridge. On the extreme left the new intake of the Shagou He and Heishi He derivatory canals is indicated, while at the extreme right at the bottom we see the new sluice gates of the Zouma He derivatory canal. (Photo. Joseph Needham, 1958.)

Dredge out the river's stones
And pile them on the embankments,
Cut masonry to form 'fish snouts',
Place in position the 'sheep-folds',*
Arrange rightly the spillways,
Maintain the overflow pipes in the small dams.
Let the (bamboo) baskets be tightly woven,
Let the stones be packed firmly within them.

---

* These are cylinder shaped 'gabions' made of parallel wooden slats some 3 to 6 metres long, filled with stones about the size of hen's eggs and sunk in the water.

Fig. 475. The Min River and Guanxian irrigation system head-works; a view looking upstream taken from the Jade Rampart Hill. In the background are the heights of the Balang Shan mountains. In the middle distance is the Han-family island, and on the right the Thousand-foot Dyke; then the main division-head (the Fish Snout) and the suspension bridge. In the foreground we can see the Diamond Dyke separating the Nei Jiang on the right from the Wai Jiang on the left, cut through by the 'water-level adjusting spillway'. On the right the roofs of the temples of Li Erlang and Lao Zi can be seen among the trees. (Photo. from E. Boerschmann (1911).)

Divide (the waters) in the four-to-six proportion,
Standardise the levels of high and low water
By the marks made on the measuring scales;
And to obviate floods and all disasters
Year by year dredge out the bottom
Till the iron bars clearly appear.
Respect the ancient system
And do not lightly modify it.

Truly an epitome of the hydraulic art. Perhaps one need only add that the benefit which the Sichuanese received from the Guanxian works was not

Fig. 476. The Cornucopia Channel, looking upstream from the terrace of Li Bing's temple on the Lidui Hill. The Phoenix Nest Cliff towers on the right; on the left the Flying Sands spillway is strongly overflowing. (Photo. Joseph Needham, 1958.)

limited to irrigation and flood protection. A stone tablet of the Yuan period (AD 1271 to 1368) records that 'waterwheels for hulling and grinding rice, and for spinning and weaving machinery, to the number of tens of thousands, were established along the canals (in the plain of Chengdu), and operated throughout the four seasons'. Thus the economic life of a whole Chinese province at the time of Villard de Honencourt and Roger Bacon depended on those noble works of civil engineering among the misty mountains of western Sichuan.

One further point before leaving Guanxian, though it somewhat oversteps the boundaries of engineering as such. The Chinese were never content to regard notable works of great benefit to the community from a purely utilitarian point of view. With their characteristic ability to raise the highest secular level to a spiritual dimension, they built on top of the 'Separated Hill' a temple to commemorate Li Bing's heroic virtue. Further back, in a scarcely less beautiful situation, on the wooded hillside downstream from the suspension bridge, another temple was raised to his son Li Erlang; this also contained living quarters for engineers, its beauty in no way diminished,

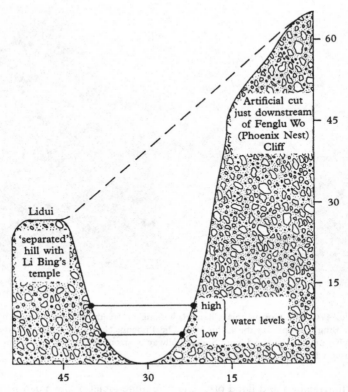

Fig. 477. Schematic cross-section of the Cornucopia Channel (data collected in 1958). Scales in metres.

while a smaller temple dedicated to Yu the Great housed the Water Conservancy Bureau. Overlaid though this may formerly have been by superstition and ignorance among the mass of people, the cult at Guanxian seemed to demonstrate one of the most attractive aspects of Chinese culture, namely that welding of Confucianism and Daoism in which, whatever may be thought of gods and spirits, divine honours were certainly owing, and paid, to the great benefactors of humanity.

The Kunming reservoirs (Yuan) and the Shandan system (Ming)

As a postscript to the story of Guanxian, mention will be made of two other projects which, though not so spectacular, greatly benefited smaller areas of territory. One is a reservoir system; the other a derived irrigation canal starting from its valley of origin through a saddle in the mountains.

First let us consider the irrigation works of the Kunming plain in Yunnan. This plain or basin surrounded by rolling uplands is centred on the provincial capital and the Gunming Lake, with its western hills crowned with woods and temples. The essential problems here were first the assurance of free passage for the waters of the lake, which were otherwise liable to flood large areas, and secondly the formation of reservoirs and artificial canals so as to distribute as widely as possible the waters of the six small rivers which feed into the lake. All this was achieved by the Yuan governor of the province, a Persian or Arab by origin, Sa'īd Ajall Shams-al Dīn (Saidianchi Shansiding), in collaboration with a local engineer Zhang Lidao, and building upon minor works previously carried out by the former indigenous and independent Tibeto-Burmese dynasty of Nan Zhao, which the Mongols had conquered in AD 1253.

Sa'īd attached himself to Genghis Kan during the Mongol expeditions in the west, and soon rose to many positions of importance under the Yuan regime, becoming governor of Yunnan in AD 1274. Here he greatly exerted himself to raise the cultural level of a backward population, erecting Confucian temples and Muslim mosques impartially, and his benevolent administration was commemorated by a stele. A dozen reservoirs with dams in the hills, more than forty sluices and a network of dyked canals lined with beautiful trees still remain in operation.

The second system of works is the White Rock Cliff (Baishi Ya) irrigation scheme, a contour canal which formerly watered a great area of fertile land (enough for more than 1000 farms) between the mountains and the desert near Shandan in Gansu province. A traveller up the Old Silk Road northwestwards from Lanzhou will find the desert lies to the right, while to the left is the glittering snow-capped chain of the Qilian mountains. For 329 kilometres or more in each direction from Shandan, the road passes through steppe country or desert scrub, crossing a great number of watercourses which take their origin in the mountains and run down to lose themselves in the sands of the Gobi. Water and its retention must have been the great problem here for centuries.

Much of what we know about the White Rock Cliff scheme is summarised on a commemorative stele erected in AD 1503 by a Daoist hermit Wang Qintie. The project was a bold one, for it tapped the Datong He River at a precipitous place (the White Rock Cliff), more than 100 kilometres to the south-west of the Shandan plain. To appreciate this, one must realise that the Datong He flows in a deep valley behind the first range of the Qilian mountains, running south-eastwards parallel with the Old Silk Road, to join the Yellow River above Lanzhou. The division-head must have been constructed at an altitude of 3800 metres, for a pass of this height had to be crossed by the canal before it could descend to the Shandan plain (itself mostly about 1800 metres), and the canal must have followed the high contour for a long way before beginning its descent. When the works were first constructed we do not know, but by the end of the fifteenth century AD they had silted up and ceased to function; repairs took three years to restore everything completely. Mr Wang, the hermit, ended his inscription with a poem:

> The White Cliff towers nine hundred metres high,
> For a hundred years the desert laid waste,
> But then came a lover of the people
> Calling for a mighty engineer,
> And Li (Bing) and Yu (the Great) lived again,
> Building new dams, new dykes.
> The rolling waters, how everyone longingly wished for them!
> Drawn forth in curving course, like the Han river, they came.
> And the men of the three Armies could settle on the plain, and
> plough.

But in after years the works again fell into disrepair, and now there is little trace of them in the neighbourhood of Shandan.

The 'Magic Transport Canal' (Qin and Tang)

The Magic Transport Canal (Ling Qu) was a work of quite different character, not primarily meant to serve irrigation but the need for freight communication across one of the principal ranges of high mountainous country which separate the north and centre from the south. It connected two rivers in Guangxi, one flowing northwards and one southwards, so that through transport was made possible between the Yangtze, the Dongting Lake and the West River flowing down to Canton.

Though the Historical Records does not mention it by name, it does give accounts of its construction, which show that its primary purpose was to keep up a flow of water-borne supplies to the armies which, in 219 BC, were sent to conquer the people of Yue. It says that Qin Shi Huang Di sent commanders 'to lead forces of fighting men on boats with deck-castles to the south to conquer the hundred tribes of the Yue. He also ordered the Superintendent (Shi Lu) to cut a canal so that supplies of grain could be sent forward far into the region of the Yue'. But this may not be the oldest reference to the canal, for *The Book of (the Prince of) Huainan* has a page or two on the First Emperor's campaigns, and gives the above statement about Shi Lu in almost the same words. They go back, then, to 120 BC, just a century after the work itself. Again, in a biography of an official concerned around 135 BC with grain transport, it is mentioned that 'elderly gentlemen said' that the Magic Canal had been dug by Shi Lu in Huang Di's time. Thus the date of the project may be considered as firmly established.

To understand what kind of work it was (and still is), we must glance at Fig. 478. The northward flowing river is the Xiang; the southward one is the Li; and between them is a saddle of hills which gave Shi Lu the opportunity for the first of all contour canals. The part of the Magic Canal which justifies this designation was called the Nan, and it branched off from the Xiang River to run alongside at a suitable level or slightly falling contour for about 5 kilometres till it met the upper waters of the Li River. These themselves had to be canalised for another 28 kilometres downstream as far as the junction with the Gui Jiang river. Meanwhile, in the other valley, a lateral transport canal some 2.4 kilometres long – the Bei Qu – was dug at a more even gradient than that of the untrained Xiang.

How the works were constructed may be appreciated from Fig. 479. A division-head called a 'Spade-Snout', reminiscent of those used at Guanxian, was built in the middle of the Xiang River and backed by two spillways discharging into the river's old bed. The embankment or retaining wall of the canal was provided with several further spillways as it wandered through the town of Xing'an, and was crossed by several bridges. In this way, a pool was formed on the same level as the connecting canal itself, through which the barges could pass when they had been worked up that far. As the division-head was placed in the Xiang River, considerably the larger of the two, most of the water in the canal (nearly 1 metre deep and some 4.5 metres wide) was derived from it. The local saying was that three-tenths of the Xiang water went into the southern or connecting end (Nan canal) while seven tenths flowed down the northern lateral one (the Bei canal). All this was in working order, and much used, during the Early Han period, especially for naval purposes between 140 and 87 BC (with a peak traffic about 111 BC) when Han Wu Di was campaigning in the south for the final reduction of Yue. Another heavy duty period came in about AD 40 in connection with the important expedition against Annam, and it is recorded that on this occasion the commander-in-chief Ma Yuan extended the canalisation of Xiang. This must mean that he improved its navigation for some distance downstream.

There is an extensive classical account by Zhou Qufei in his *Ling Wai Dai Da* (Information on What is Beyond the Passes) of AD 1178. After a general description of the Canal, he writes:

> Passengers travelling along at certain points are sometimes scared
> out of their wits, for about 2 *li* (about 1 kilometre) from the intake
> where the 'spade-snout' divides the waters and makes one branch
> enter the embanked canal, there is another spillway (which lets off

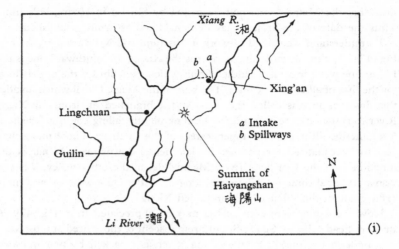

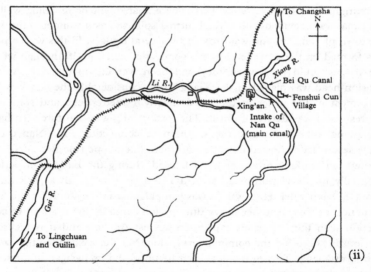

Fig. 478 Two sketch-maps of the geography of the Magic Transport Canal in north-eastern Guangxi, first built in 215 BC. This is the most ancient of all contour transport canals; it joins a northward flowing river, the Xiang, with a southward flowing one, the Li (called the Gui lower down), by means of a canal cut, and is constructed for some 32 kilometres through a saddle in the high hills. (i) The canal begins (*a*) near the little town of Xing'an, and runs for 5 kilometres westwards (*b*) to join the uppermost waters of the Li River, which had to be trained and canalised for a further 28 kilometres. (ii) This shows, on a larger scale, the lateral canal dug to give a more even gradient without rapids on the east bank of the Xiang River. The division-head in the pool above it gave its name to the village of Fenshui Cun. The modern railway crosses the ancient canal and then runs along south of the canalised Li River.

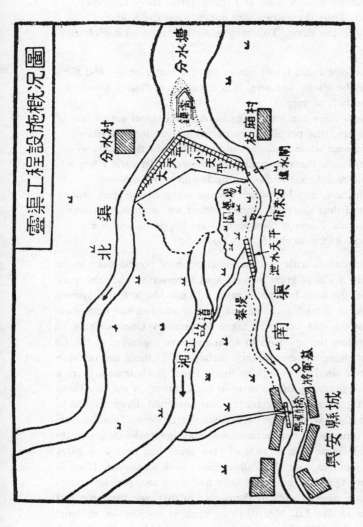

Fig. 479. Plan of the head-works of the Magic Transport Canal. North is towards the top left-hand corner. The Northern Canal brought the barges up into the Fenshui Tang pool, just below which the Xiang Jiang water was made to flow in two directions by the division-head or Spade Snout. Two spillways flanked this on either side, dismissing overflow water into the old bed of the Xiang River lower down. Passing now along the South, or main, Canal, the vessels entered a flash-lock gate and proceeded through the small town of Xing'an westwards. Throughout its length the canal is built upon, and protected by, massive embankments and retaining walls of dressed stone. The water-level is moreover controlled by a further series of spillways on the north-eastern side, first the Feilai Shi, 'the stones that came flying of themselves'; then the Yishui Tianping (Water-emitting Balance) and so on.

excess water). Without this spillway, the raging force of the spring freshets could damage the retaining wall and the water would never reach the south. But by its aid, the violence of the water is abated, the embankment is unbroken, and the water in the canal flows smoothly on. Thus the extra water drawn from the Xiang penetrates into the Rong. This may really be called an ingenious device.

The depth of the canal is not (more than) a metre or so, and the breadth may be about 6 metres; it is adequate to float a vessel of 1000 *hu* (bushels of cargo).

In the canal there are 36 lock-gates. As each vessel enters one of these lock-gates, (the people) immediately restore it to its closed position and wait while water accumulates (within the lock), so that by this means the ship gradually progresses. In such a way they are able to follow the mountainside and move upwards.

On the descent, it is like water flowing down the stepped groove of a roof, and thus there is communication for the boats between north and south. I myself have seen (I am happy to say) the historic traces of the work of (Shi) Lu.

This eye-witness account, with its clear description of pound-locks in the late twelfth century AD, is of great interest, and necessarily raises the question of the time of the introduction of gates into the Magic Canal system. Zhou Qufei speaks as though locks of one kind or another had been there since the Qin, but this can hardly be taken for granted. One can have no difficulty in attributing the spillways to an engineer so ingenious as Shi Lu must have been, a younger contemporary, perhaps, of Li Bing, and certainly of his son, who had used them with so much effect at Guanxian; there is moreover the close similarity of the division-head 'snout' in the two cases. But whether Shi Lu installed any gates remains uncertain. Evidence will be presented later (p. 246) which shows that sluice-gates were a familiar technique in the first century BC, and references at that time make clear that the Canal of the Wild Geese (see p. 193) of Han times had flash-lock gates, especially at the junction with the Yellow River near Rongyang. If so, in Warring States and Qin this canal may have had them too, perhaps indeed it could hardly have worked without them, and in that case they may well have been known to Shi Lu. Nevertheless, there is no positive evidence from texts or elsewhere that he installed them in the Magic Canal or its approaches.

If he did not, then one has to picture southbound barges being towed up the trained Xiang River by gangs of trackers into the Bei Qu section, and so into the pool by the great spillway. Having passed the division-head, they would reverse into the canal, where there was no rapid current, for the level course had been made as winding as possible precisely in order to slow it. No reverse was needed at the other end, where the boats would glide fairly fast down the trained Li.

The first information we have of gates comes from a period still relatively early, the Tang, and concerns the important restoration of the canal carried out in AD 825 by Li Po. In the *Commentary on the Waterways Classic* we read:

> During the Tang dynasty, at the beginning of Baoli reign period (the banks of the canal) were collapsed and broken so that boats could not get through. The Inspector-General Li Po therefore caused stones to be piled up in courses to make a dyke like a 'spade snout' to split and divide (the Xiang River) into two streams. In each stream there were set up stone (abutments for) flash-lock gates (lit. dipping gates), with one keeper in charge (of each), who let people freely open and shut (as required). When the Li River (gate or gates, was or were) closed all water flowed into the Gui River (to the south); when the (gate or gates, was or were) closed on the Gui side, then all (the water) returned to the Xiang River (northwards). And moreover at the Xiang River he cut a 'dividing-water canal' 35 paces long (54 metres), in order to facilitate the movement of ships.

Taking this last improvement first, the most obvious suggestion is that Li Bo isolated the division-head in the form of an island quay, making a canal behind it along the top of the dam so that it was no longer necessary for the boats to reverse. But the flash-locks were no less important. From other sources we know that there were eighteen of them, first set up in rough construction in wood, but then greatly strengthened by Yu Mengwei in the further improvements that he carried out in AD 868. Their exact positions we do not know, but from the text it seems sure that at least one of them was at the Li end of the canal, and at least one other at the Xiang end. There was more need for them, however, on the graded approaches rather than on the level canal, so it is probable that the majority were erected in the actual course of the Li, the Bei Qu section parallel with the Xiang, and in the Xiang's course lower down as well. From the ninth century AD at the latest onward, barges ascending on either side were hauled through flash-locks, probably with winches, and towed by a much reduced company of trackers on the relatively level intermediate sections. Then comes the information that the number of gates was no longer eighteen but thirty-six. This is still the approximate number today.

On the evidence as a whole, it seems most likely that the introduction of pound-locks into the Magic Canal system occurred in the tenth or eleventh centuries AD; as we shall see later on, this fits in with the rest of the Chinese evidence extremely well.

The significance of the Magic Canal as a link in a chain of communications, altogether extraordinary for the third century BC in any part of the world, should not be overlooked. By navigation of the lower Yellow River, the Canal of the Wild Geese and the Han-Country Conduit, the Yangtze, the Xiang River leading south from the Dongting Lake, the Magic Canal and the West River, the first Han emperor in 200 BC found himself in possession of a single trunk waterway extending from the 40th to the 22nd parallel of latitude, that is to say a distance of more than 2000 kilometres in a direct line, and doubtless more than double that as the vessels sailed. And lastly the Magic Canal resembles the Guanxian in that, although it is a work of the third century BC, it has been repaired and set in order once again in our time, and continues to carry out heavy duties. Few if any other civilisations could demonstrate a work of hydraulic art in continuous use for well over 2000 years.

The Grand Canal (Sui and Yuan)

Our remarks about the Grand Canal as a whole can be brief because several of its precursor or component sections have already been mentioned. The work of combining these together was done by the Sui (AD 581 to 618) when the need to link the capital at Luoyang with the key economic area of the lower Yangtze valley became imperative. In the last decades of the thirteenth century AD, under the Yuan emperors, the same need continued, but as the capital was now at Beijing a vast remodelling was carried out. Finally, it formed a continuous waterway following the 118th meridian in an S-shaped course from Hangzhou in the south to the furthest northern parts of the North China Plain. Covering 10° in latitude, it would be comparable with a broad canal extending from New York to Florida, its total length attaining some 1770 kilometres. Its summit, reached when skirting the mountains of Shandong, was some 42 metres above sea-level.

The oldest pioneer section was the Hong Gou, later known as the Bian Qu Canal, which connected the Yellow River with the Huai Valley. As we have seen, this was at least as old as the fourth century BC, and experienced naval activities as well as civilian transport. By AD 600 it had silted up so much that Yuwen Kai, the chief engineer of the Sui, decided on an entirely different alignment. Broadly speaking, this ran parallel with the Hong Gou but south-west of it. After diverging from Chenliu ((9) in Fig. 464), it passed Suzhou (120) not Xuzhou (10) and joined the Huai River directly, west of Hongze Lake, without making use of the Si River. Its length was some 1000 kilometres. In the north-west, from near Kaifeng and the junction of the Yellow River at Sishui (just downstream from the entry of the

Luo River at Luoyang), the course was much the same as the Hong Gou, but special works were erected to protect its opening. In AD 587, Liang Rui, another eminent Sui engineer, built a massive westward continuation of the 'Metal Dyke' (see p. 184) on the south bank of the Yellow River; this contained lock-gates for regulating water levels and double-slipways over which boats could be hauled when the differences were too great for flash-lock operation. The main project, the new Bian Qu Canal, was completed in AD 605 using more than five million men and women. Throughout the Tang and the Song (seventh to thirteenth centuries AD) the New Bian, the 'Grand Canal' of the Sui, continued in active, indeed heavy, use.

The realities of life on the canal in those times can be gathered from the diaries and memoirs of foreigners like the Japanese monk Ennin, whom we have met before. In AD 838 he journeyed from the coast of Yangzhou on one of the lateral feeder canals from the numerous salt works which had been built or repaired between San Guo to Sui times. On a canal which was straight as far as the eye could see, a train of forty boats, many lashed two or three abeam, was pulled slowly but efficiently by two water buffaloes. Once, a bank caved in, but Ennin's party got through by digging, and when they came near the Grand Canal, the salt boats, three to five lashed abeam, followed after one another continuously kilometre after kilometre.

The engineers of the Tang introduced some modifications of the system but no fundamental changes. In AD 689 a branch was built to go from Chenliu north-east to the trunk waterway in Yanzhou in Shandong. Then, between AD 734 and 737, another new canal was constructed by Qi Huan, which skirted the north shore of the Hongze Lake and brought traffic directly from the New Bian Canal to Huaiyin, thus short-circuiting the dangerous rapids of the Huai River. At the same time, he also made a short-cut near Yangzhou which saved some 20 kilometres.

The portion of the Grand Canal south of the Yangtze was not an entirely new enterprise of the Sui because there had been earlier artificial waterways in the region (see p. 196), but it took a new course, completed in AD 610. Some 400 kilometres in length, it skirted the eastern side of Taihu Lake and put Hangzhou in direct communication with the north, thus enabling the supplies and products of the south-eastern coastal regions to flow to the capital. In order to ensure traffic between the Yellow River and Chang'an, Yuwen Kai restored the old canal in Shaanxi which had been dug in 133 BC.

Here we may pause for a moment to note a rather intensive development of urban waterways in the Venetian style in these cities, so useful as part of the traffic system. Many of these can be seen in Suzhou today, encircled as it is by the Grand Canal (Fig. 480), and one can find the remains of an elaborate network in Hangzhou.

As for the portion of the Grand Canal north of the Huang He River,

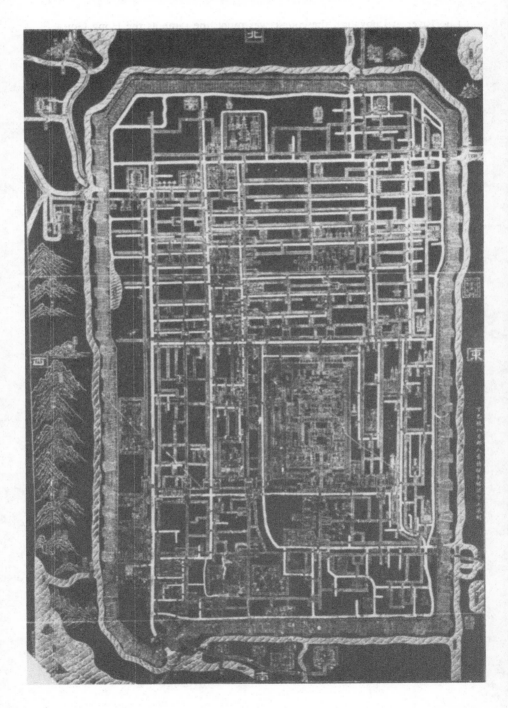

about 997 kilometres in length, this was truly a new enterprise of the Sui. It took advantage of a short river, the Qin, descending southwards from the Shanxi mountains and falling into the left of the Huang He a little east of Sishui, thus rendering the lower reaches of this navigable. It then struck across country by a short branch to join the head-waters of the canalised Wei River, which flowed north-east to the neighbourhood of Tianjin.

During the Tang and Song dynasties, the highest officers of state occupied themselves with the engineering problems involved in conserving the Bian Canal as the great artery of tax-grain transport from south to north-west. Fortunately, in the Five Dynasties interim which followed the collapse of the Tang and before unification once more in the Song, though this economic area was split up into different political units, there was little or no destruction of the works themselves. After AD 960 the Song intensified the care of the waterway as part of their expanded water conservancy programme, which outdistanced that of any previous dynasty. Their capital was at Kaifeng, but by the twelfth century their hold on it was becoming precarious, and, after the collapse of its defences against the Jin Tartars in 1126, the Song general staff deliberately broke the dykes south of the Yellow River, thus destroying most of the works of the Bian Canal. Indeed, it must have been in a fairly dilapidated condition when the Yuan (Mongol) overwhelmed Jin, Liao and Song alike, and obliged historians to count the years of their complete dynastic power from 1280.

Since the Yuan dynasty contained so much northern territory which was outside the bounds of China proper, Beijing was the natural choice for the capital. In spite of this, the long-established pattern of Chinese social and

---

Fig. 480. Map of the city of Suzhou carved on stone in AD 1229 by Lu Yan and two other local cartographers. North is at the top. This stele is now preserved in the old College (now a Middle School) attached to the Confucian Temple in the lower left (south-west) corner of the city. The Grand Canal comes in at the top left-hand corner as the third stream down, then it flows past the western city wall and out at the bottom, just beside the large Chinese character (*Nan* = South). Other waterways surround the city, providing canals within it almost as numerous as the streets, crossed by 272 bridges. Marco Polo, who may well have seen the stele itself, referred to 6000 bridges, but his scribe may have exaggerated by two powers of ten. Part of Lake Taihu is visible in the extreme left-hand bottom corner; above it there are the hills which protect the city from the lake, and at the top left-hand corner the famous hill of Huqiu Shan is shown. The Buddhist temple of Baoen Di can be made out in an enclosure at the extreme north of the city, and near the centre, above the inner walled complex of government buildings labelled Pingjiang (the old name of Suzhou), is the great Daoist temple Xuanmiao Guan. (Photo. from Chavannes (1913).)

economic life persisted unchanged, simply with the Mongols (and other foreigners) at the top, but it did mean that Beijing could not become a true administrative centre without streamlining the tax-grain system. To achieve this, the essential thing was to make a short-cut in such a way as to bring the east-central economic area more directly in touch with the north. From Hangzhou to the Huai River no great change was needed, but north of the Huai it was necessary to abandon the westward trend towards the age-old hub of Rongyang near Kaifeng and to plan a more direct route. The line of the old Bian Canal was thus entirely replaced by a more easterly one, which made its way over the shoulder of the Shandong mountains and crossed the Yellow River far to the east of the former junction points.

This new alignment was truly a summit canal in a way which the Bian Canal had never been, and doubtless its planning drew some inspiration from the known success of the Magic Canal. Though sea transport now began to provide serious competition, the Yuan dynasty still relied largely on the waterways, and this brought about a remodelling of the Grand Canal which has lasted throughout the Ming and Qing periods down to the present day. Who, after all, could fail to appreciate the political and economic consequences of so bold a conception – an artificial river running north and south in a country where most of the natural rivers run from west to east?

The northernmost part, the Tong Hui He (Channel of Communicating Grace) was comparatively short, running 82 kilometres from Beijing to Tongzhou, and was completed in AD 1293 by the astronomer Guo Shoujing, who was also a brilliant civil engineer. This section needed about twenty lock-gates, and Marco Polo wrote glowingly about it. In AD 1307, Rashīd al Dīn al Hamadānī described not only its completion but also the whole Grand Canal system, mentioning especially its flash-lock or pound-lock gates, its capstan slipways and the protection of the trees planted alongside. Much later, in 1558, a special monograph, the *Tong Hui He Zhi* (Record of the Canal of Communicating Grace) was written by Wu Zhong, and within a dozen years it was translated into Persian. Thus, those who knew what had happened were much impressed by what Chinese engineering could do.

The second part, (*b*) in Table 58 (p. 224), running from Tongzhou to a point near modern Tianjin had originally belonged to the Sui system but was greatly improved in Guo's time and was called the Bai He (White River). The third section (*c*) also followed the Sui route, but about halfway along it, at Linqing (124), an entirely new canal, the Union Link Channel (Hui Tong He) was thrown off southwards to cross the northern course of the Yellow River at right angles. This had thirty-one locks in a distance of about 125 kilometres.

The important summit section (*f, g*) was the work of the Mongol engineer Oqruqči in AD 1238 and the following years using plans drawn up by Guo.

Known as the Ji Zhou He, it connected Jining with the Qing Ji Du canal, and so with the sea, by means of the river north-eastwards, a stretch of more than 70 kilometres. On the other side, it linked up with another work of the Jin period, the Huan Gong Gou, which had first been built in AD 369.

In this summit section the water attains a height of 42 metres above the mean level of the Yangtze at the junction point. As Oqruqči left it, the water supply for the highest levels (always the greatest problem of summit canals) was unsatisfactory – so much so indeed that throughout the Yuan period the canal could not always compete with the alternative advantages of sea transport. Necessary remodelling of sections (*d*) to (*i*) to overcome this was carried out by the Ming engineer Song Li in AD 1411; this brought the most difficult part of the Grand Canal to a high level of efficiency. With the help of Bai Ying, Song solved his problem by using the waters of the Wen and Guang Rivers more effectively; this involved constructing a new 1.6 kilometre-long dam north of Ningyang to form a large reservoir which would always keep the canal full with the aid of a forking lateral canal from the Wen. These major works were completed in 200 days by a force of 165 000 men. Song also inserted four smaller reservoirs near the canal itself under the name of 'water boxes' with sluice-gates to let go any excess water. These water boxes were, it seems, something like the side ponds of modern pound-locks (see p. 246).

However, by AD 1327, the Grand Canal had attained definitive form, extending in all approximately 2575 kilometres (see Table 58), with a dia-grammatic profile as shown in Fig. 481. It may truly be said that the work of Guo Shoujing and Oqruqči in 1238 constituted the oldest successful fully artificial summit canal in any civilisation. Yet we should remember that the installation of the pound-locks in the approaches to the Magic Canal some time between AD 900 and 1170 had, in a sense, converted that system into one of the summit kind. Nevertheless, the great work of engineering which constitutes the Grand Canal (Fig. 482) is all the more remarkable because it had to connect two of the greatest rivers in the world and one of them the most changeable.

The Qiantang sea-wall (Han, Wu Dai and Song)

Lastly, a few words about a work of rather different character to any so far mentioned, namely the sea-wall which gives its name to the Qiantang estuary (Fig. 483). It is upon this 80 kilometre-long inlet that Hangzhou, the southern terminus of the Grand Canal, is situated, and through which the waters of the Fuchun Jiang (Happy Spring River) flow to the sea. The sea-wall was, in a way, an embankment like all other embankments, but it had to face particularly violent attacks due to both the serious storms for which the bay is famous and to a well known tidal bore.

Table 58. Details of the Grand Canal in its final form (late Qing).[a]

Section	Kilometres (cumulative)	Intervals (kilometres)	Elevations above Yangtze (mean water level)		Depth of canal (metres)
			Surrounding ground (metres)	Water level (metres)	
Beijing (Tong Hui He)	0	29	36	34	3
Tongzhou (Bai He)	29	124	28	26	3–8
Tianjing (Yu He)	153	386	8	7	3–10
Linqing (Hui Tong He)	539	113	36	35	3
Yellow River, north side	652	—	38[b]	35	c. 9[c]
Yellow River, south side (Hui Tong He, Qing Ji Du and Ji Zhou He)	652	69	38[b]	42[d]	
Nan-wang Zhen[1] summit (Ji Zhou He)	721	21	52	42[e]	4–4.3
Jining[2] (Huan Gong Gou)	742	143	40	35	4–7
Linjia Ba[3] (south end of lakes today) (Huan Gong Gou)	885	250	36	35	3–10

Huaiyin (Shanyang Yundao)[1]	1135	186	18	16	4–8
Yangtze River, north side[2]	1321	–	5	0	12–15
Yangtze River, south side (Jiang Nan He)[3]	1361	47	5	0	4
Danyang[4] (Jiang Nan He)	1408	298	17	2[e]	4
Hangzhou	1706		4	0	
		1666			

[a] Prepared in conjunction with A. W. Skempton.
[b] The ground level, not that of the top of the dyke.
[c] Very variable.
[d] Canal running on an embankment. This is not shown in Fig. 481.
[e] Canal in cutting.

[1] 南旺集　[2] 清峯　[3] 蘭家埃　[4] 丹陽

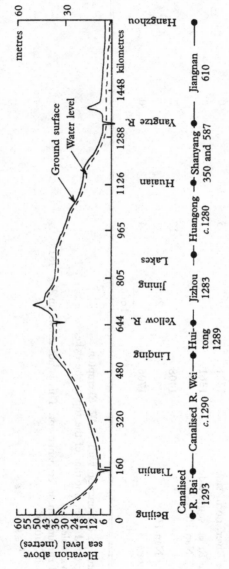

Fig. 481. Diagrammatic profile of the Grand Canal (with A. W. Skempton, after Zheng Zhaojing and others). All dates are AD.

Fig. 482. An engraving of the Grand Canal running along the edge of the Gaoyu Lake, but separated from it by an embankment to assist traffic in bad weather. This shows how J. Nieuhoff's embassy to China in 1655 to 1657 saw in 1656. 'In former times', he wrote, 'all vessels coming from Nanjing and the Yangtze, and bound for Beijing, were obliged to wait below the walls of this city (Gaoyu) in stormy or foggy weather. But as these delays were very vexatious for trade, it was thought good, to avoid the perils of the lake, to build along its eastern side a canal 60 stadia long; and this is done with square white dressed stone blocks of such a size that one cannot imagine where they were got from, seeing that in the neighbouring provinces there are no great stone hills or quarries.' Note the cross-sections of the dykes indicated by the artist in the foreground. (Photo. from Nieuhoff (1665).).

Fig. 483. The Qiantang sea-wall near Haining. Near this spot stands a hexagonal six-storeyed pagoda dating from the Qianlong reign period (AD 1711 to 1799) and 'a platform for watching the tidal bore'. In the foreground can be seen one of the remaining cast-iron geomantic bulls placed along the wall in 1730. (Photo, Joseph Needham, 1964.)

The first beginnings of a dyke, which probably only protected the settlement from which Hangzhou developed, are associated with Hua Xin, the governor of the region between AD 84 and 87. As centuries went by, the seawall gradually grew, and in 822 the poet Bai Juyi, the then governor, extended it to protect the region's irrigation systems.

Originally, it was composed of bamboo gabions filled with stones with large wood baulks set in amongst them, but in AD 1014 parts of the wall were made of earth bonded with straw. However, some twenty years later, piled stones were used for facing the wall. There seems to have been much reluctance to undertake the very large task of cutting stone facing throughout, and it is clear that three alternatives presented themselves. First, filled bamboo gabions could be used after the manner of Guanxian, anchoring them with wooden piles and binding them with iron chains. Secondly, it would be possible to build the embankment of tamped and reinforced earth such as was used for dams and weirs inland; but though less expensive it would not last as long. Finally, in AD 1368, a wall of stone rubble was built, but this had the failing that for a long time it was not backed by an earthfill embankment of sufficient size; this fault was remedied in 1448 when

Yang Xuan introduced the practice of building the masonry in steplike fashion, the better to break the force of the waves. Then later, in 1542, Huang Guangsheng made a special study of bonding which he described in his *Tang Shuo* (Discourse on Sea Walls), and this proved effective down to modern times. In addition, the use of iron clamps for the blocks of stone, the building of breakwaters, groynes and piers out into the sea, and drainage canals materially assisted the success of the project. Today it weaves its way along both sides of the estuary for over 300 kilometres, its masonry rising to an average of almost 10 metres above low-water level, and efficiently performs its ancient task.

THE LITERATURE ON CIVIL ENGINEERING AND WATER CONSERVANCY

From all which has been said, it will have been noticed that the chief sources of information on hydraulic engineering in ancient and early medieval times in China are the dynastic histories. The Historical Records of about 90 BC and the History of the Former Han Dynasty of about AD 100 contain long chapters on canals and waterworks, while some of the later ones, such as the *Song Shi* (History of the Song Dynasty) of AD 1345, contain much data yet to be analysed fully. But few fragments from the Han about, for example, water conservancy survive, and our knowledge of their ideas and practice comes from recorded speeches and memorials. Most of the technical literature dates from the fourteenth century AD and later.

It is hard to make any sharp distinction between 'hydrographic' treatises and books specifically devoted to hydraulic engineering techniques. They usually overlap, though the latter are commoner in the latest periods. The oldest hydrographic survey which has come down to us is the *Shui Jing* (Waterways Classic), attributed to Sang Qin of the Former Han, but more probably compiled in the third century AD. It was greatly enlarged at the end of the fifth century AD or the beginning of the sixth by the geographer Li Daoyuan, who entitled it the Waterways Classic Commented. But it is mainly geographical and only tells us about canals, though it was much commented upon by other scholars.

From the tenth to the thirteenth centuries AD, under the Song dynasty, quite a number of relevant books were written, many of which survive. Two of the more important are the *Wu Zhong Shui Li Shu* (Water Conservancy of the Wu District) of 1059, the result of many years' study of the canals of Jiangxi, and the *Si Ming Tuoshan Sui Li Bei Lan* (Irrigation Canals of the Mount Tuo District) of 1242, which is a historical account of their development in the neighbourhood of Ningbo. These and other works mark a period exceptionally rich in civil engineers.

The first great compendium dates from the Ming. Pan Jixun served four terms of office between AD 1522 and 1620 as Director-General of the Yellow River Works and the Grand Canal, becoming the greatest authority in his age on hydraulic engineering. His *He Fang Chuan Shu* (General View of Water Control) of 1590 includes many maps, copies of memorials, edicts and other official documents and many interesting discussions by the author, written in the form of dialogues with imaginary opponents. In spite of changing methods to fit changing circumstances, it was still considered authoritative in the late eighteenth century.

In the seventeenth century AD, the greatest hydraulic engineer was Jin Fu, who worked especially on the Yellow River and the Grand Canal. He superintended extensive dredging operations, improved the form and strength of embankments, was concerned with drainage of the lower Huai area and constructed a successful 153 kilometre canal which short-circuited a bad patch of the Yellow River. He wrote the *Zhi He Fang Lue* (Methods of River Control) in 1689, and, though it remained unpublished until 1767, it is considered of comparable importance to the treatise by Pan Jixun.

The eighteenth century was very rich in books, seeing the appearance of the monumental *Xing Shui Jin Jian* (Golden Mirror of the Flowing Waters) of 1725 by Fu Zehong, the most comprehensive treatment of all Chinese waterways, natural and artificial. Of works devoted solely to engineering techniques rather than historical description, notable was Kang Jitian's *He Qu Ji Wen* (Notes on Rivers and Canals) of 1804, the author of which was a man of great experience in the field as well as in the administrative office. It is considered to be one of the best books on the subject in the whole of Chinese literature.

TECHNIQUES OF HYDRAULIC ENGINEERING

Planning, calculation and survey

It is obvious that works on the scale described could never have been undertaken without a considerable staff of ingenious planners. Unfortunately, we have few indications of exactly how the planning problems presented themselves to the men of past ages, and how far they were able to get in solving them without having to fall back on guesswork. There is a passage in the *History of the Former Han Dynasty*, however, which reveals something of the situation behind the scenes. A problem in 30 BC was to try to relieve the pressure on the lower Yellow River dykes by building canals, and one important cause of trouble was the silting up of the Tunshi River, one of the channels of the delta. The upshot was a realisation of the necessity of associating good mathematicians and engineers with the administrators in water conservancy and control works. This occurred because, by themselves, the administrators were capable of making gross miscalculations and costly mistakes, yet, interestingly, the passage reveals some conflict between the 'back-room boys' and the regular officials. It is also important because it gives the oldest description of the bamboo gabions filled with stones which played so important a part throughout Chinese hydraulic engineering. That they were unfamiliar in about 28 BC, when two men seemed to claim their invention, may mean that they were really introduced about this time.

One would naturally expect that specimen calculations about canals and dykes would appear in the mathematical literature, and so they do. For instance, the Han *Jiu Zhang Suan Shu* (Nine Chapters on the Mathematical Art) of the first century AD contains many problems on dyke building, giving results in material, labour, time and so on.

Again, in discussing Chinese geography (volume 2 of this abridgement), reference was made to the rectangular grid system in map making. Plumb-lines and the *groma* (a device with four plumb-lines hung from a wooden cross for determining right angles), chains, cords, graduated poles and water-levels with floating sights were early in use, while in AD 1070 sighting tubes mounted in various ways formed proto-theodolites. The cross-staff, generally ascribed to the fourteenth century AD in Europe, was in use in eleventh century China, and compass bearings were also utilised at the same period. Moreover, as early as AD 263, Liu Hui in his *Hai Dao Suan Jing* (Sea Island Mathematical Classic) was giving many examples of the use of similar right-angled triangles. Whether the hodometer (volume 4 of this abridgement), available from Han times onwards, was actually used in practice for measuring distances, remains uncertain.

After the route of a canal had been provisionally decided, trial borings were made along it to determine the nature of the ground, as Lou Yao recorded in AD 1169 when writing about the Song Canal. Further, in order to secure adherence to the original alignments, reference marks in the shape of stone or iron plates or statues were fixed in the sides and beds of canals as a guide for periodical deepening and silt removal, as at the Guanxian division-head. A 'nilometer' in the stricter sense is the 'Two Fishes' gauge rock which has measured the Yangtze water levels since AD 763 at Fouling in Sichuan.

Drainage and tunnelling

From time to time, the engineers were faced with difficulties due to springs, underground watercourses and loose or shale-type soil which was prone to landslides. We have an example of the efforts made to cope with these things from quite an early time, namely that of Han Wu Di. Sima Qian, writing in his Historical Records, describes the building of a canal which would leave the river Luo and irrigate land lying east of Zhongquan:

As the banks were liable to slide and crumble easily, (a series of) wells was dug, the deepest of which was 122 metres; and there were wells all along at regular intervals. At the bottom they communicated with each other (by a tunnel) through which the water flowed. The water came down until it met, and flowed round, the Shangyan mountain, east of which the canal continued more than 5 kilometres until it reached the hills. This was the first time that a subterranean canal with well openings was built.

More than ten years after its completion, the water was coming through alright, but very little benefit had accrued from it for agriculture.

Though this work of Zhuang Xiongpi (if indeed he was the engineer) seems to have had only a qualified success, it must be admitted to be of considerable technical interest. One is strongly reminded of the traditional *qanāts* of Iran. These were devices to make use of mountain water sources which normally sink out of sight when they reach the foothills, losing themselves in the porous valley soil. The flow is tapped near the base of the hills and led along a subterranean channel above the impervious clay beneath it and under a succession of ventilation and excavation shafts. It then flows out into a reservoir, from which fields and settlements can be continuously supplied. The early history of the *qanāt* is obscure, but Marco Polo refers to such constructions. However, Zhuang worked during the Sassanian period (third to seventh centuries AD), when, from the upper waters of the River Diyala, a derivative canal was led southwards by a long tunnel through the Jebel Hamrīn range of hills so as to bring water to the land at the edge of the plain below. But the qanāt system never became widespread in China proper, presumably because circumstances did not call for it. Perhaps when the history of the technique is fully known, China may be found to have made a contribution.

Dredging

In considering the scouring of rivers and canals, it was altogether natural that from time to time the Chinese engineers should have attempted mechanical means as well as the use of flowing water. In AD 1073, during the Song, after devastating floods in Hubei, a Yellow River Dredging Committee was established. A candidate-official, Li Gongyi, then came forward with the suggestion for an 'iron dragon claw silt dispersing machine', namely a weighted and toothed rake. It was to be towed by two boats up and down to keep the loose bottom material on the move. Improved in design by using heavier materials, the resulting 'river deepening harrow' was a beam 2.4 metres long fitted with spikes each 0.3 metres long and sunk to its work by windlasses on two ships. Several thousand of these were made and used, but there is no informed judgement about their success.

Five centuries later, in AD 1595, towards the end of the Ming, an Imperial Censor gave his views on the best ways of keeping rivers clear, especially the Grand Canal. He recommended three methods. First, the traditional excavation of the bed, as far as possible during low-water seasons. Secondly, for 'all official and private boats coming and going towing "bed-harrowing ploughs", and sailing with the wind, scraping the bottom as they go, so that the sand has no peace to sink and settle'. This was just the scheme of Li Gongyi.

The third method was to imitate 'the hydraulic mill and the hydraulic trip hammer' and to build machines 'which use the current of the water to roll and vibrate, so that the sand is constantly stirred up and cannot accumulate'. More difficult to visualise, it suggests moored paddle-wheel vessels like ship-mills (p. 303) with boom agitating rakes worked from eccentrics, with or without connecting-rods on each side. Unfortunately, once again we do not know whether these were used and, if so, their efficacy. Yet the second method is in use down to our own times, as the illustration in Fig. 484 testifies.

Today, one sees dredging operations mainly in harbours, such as is shown in a sketch made in the 1950s by A. L. Carmona (Fig. 485); though descriptions in Chinese literature seem to be very scarce. As will be seen, a large rectangular bucket dredge on the end of a long spar strengthened and shod with iron is let down into the water alongside a barge with a large hold. The dredge is connected at the front by a cable to a pedal-operated windlass in the steersman's deckhouse at the stern, and this brings the dredge to the surface when full. It is then caught on a hook on the end of a chain hanging from one end of a lever. When the other end of this lever is hauled down, the dredge and its contents are swung inboard and emptied into the hold. Towards the middle of the eighteenth century AD, Bernard de Belidor illustrated dredgers with treadmills and just such long-shafted shovels in his *Architecture Hydraulique,* but whether they stem from a Chinese or a European original remains uncertain.

Reinforcement and repair

The need for internal bonding in structures, which ultimately led to reinforced and pre-stressed concrete, was of course felt by ancient and medieval Chinese engineers, and we have already seen examples of their methods in, for instance, the Great Wall and the Qiantang sea-wall.

Without exception, the most important material used in China for bonding was bamboo, with its remarkable tensile strength. Since it was also available in such unlimited quantities, the earliest technique was probably to

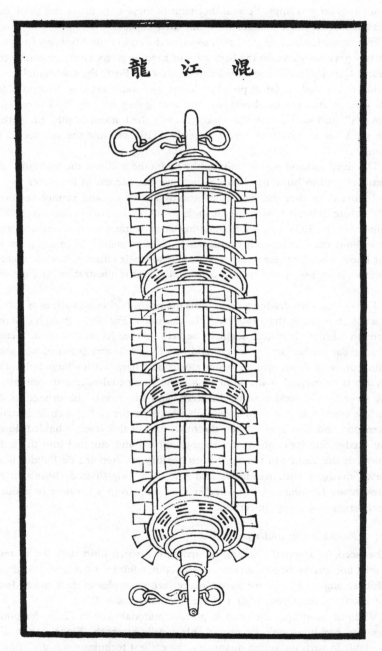

龍 江 混

Fig. 484. Towed scrape-dredge or rolling suspensifier, from the *Illustrations and Explanations of the Techniques of Water Conservancy and Civil Engineering* of AD 1836. Drawn along the bottom by a vessel proceeding upstream, it raised clouds of silt from the bed (hence its name), and so increased suspension clearance.

Fig. 485. A Chinese dredger of traditional type.

leave in position the baskets or skips in which stones or earth had been carried to the spot. Then, as time went on, the elongated gabion, a sausage-shaped openwork crate of bamboo packed with stones, was developed. This stage must have been reached by 28 BC. The great advantages of this invention (Fig. 486) have already been emphasised in connection with the Guanxian works (see pp. 202 ff.); the relative lightness of the gabions permitted their use on deposited (alluvial) subsoils without deep foundations, and their porous nature gave them a most valuable shock-absorbing function, so that surges and sudden pressures did no damage to the defences. Elongated gabions of wooden slats were also utilised.

It is interesting to find that the same device was also used in Europe, at least from the fourteenth century AD onwards, especially in the sea dykes of the Netherlands. Here, bales of compressed seaweed, or a screen of compressed seaweed within piling, or even bundles of reeds fixed down with their roots pointing seaward, were used outside the Dutch clay polder dykes. Such shock-absorbers, less resistant to decay than bamboo basket-work, had to be renewed every five years.

Besides the gabions, fascines formed of huge bundles of gaoliang stalks

Fig. 486. A weir of gabions near Chengdu in Sichuan. The plaited bamboo cylinders filled with stones can be built into any desired formation; in structures such as this, the successive layers are covered with bamboo matting. (Photo. Joseph Needham, 1943.)

fastened with bamboo ropes were also devised. These were very convenient when the water was heavily laden with silt, for solid material would quickly be deposited in the gaps through which the water filtered, and in time the whole structure would become very compact. Such brushwood fillers were a Song invention, and Fig. 487, taken from a Qing dynasty book of about AD 1775, the *Xiu Fang Suo Zhi* (Brief Memoir on Dyke Repairs), shows a drawing in Chinese style of the method of handling them. More than a century later, the traveller M. Esterer was present in 1904 at a closing of a gap in a dyke of the Yellow River. The dyke was 9 metres broad at the bottom and 3.3 metres at the top; its height was 10 metres. The gap to be filled was 11 metres wide at the bottom and more than 16 metres at the top, with water pouring through it. Gabions and fascine bundles were used, handled by 20 000 men hauling on cables 30 metres long; Fig. 488 shows a similar giant bale, in this case some 6 metres by 15 metres, stopping a breach, also on the Yellow River, but in 1935.

Graphic descriptions of the closure of gaps are given in Chinese literature. One notable example is in the *Dream Pool Essays* of Shen Gua written during the Song in AD 1086. Concerned with closing Dragon Gate gap in the Yellow

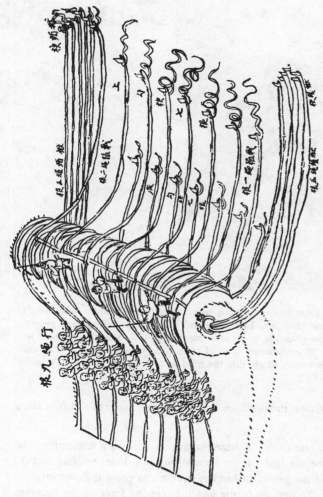

Fig. 487. Manhandling a giant fascine bundle into position: a sketch in traditional style from the *Brief Memoir on Dyke Repairs* of Li Shilu (c. AD 1775). The nine hauling ropes, with the hauliers, are seen on the left, the side towards the water. The five 'end-head bundle cables' pass through the centre of the fascine, and, being slowly paid out, act as brakes. The seven 'overhook cables', continuous with the seven 'underhook cables', form a safety cage in which the fascine bundle can roll; they would have to be re-pegged from time to time.

Fig. 488. A great bale of gaoliang stalks being lowered into a dyke breach on the Yellow River near Liaocheng and Donga. The cordage and pegging of these stacks gives life to the drawing in Fig. 487. When such a bale is set in place, 30 centimetres or so of earth is deposited on top of it so that it settles down more firmly into the mud and silt of the bottom. (Photo. E. Todd, 1935.)

River at Shanghu, near the northern capital, use was made of a fascine about 91 metres long:

> Gao Chao, one of the assistant engineers, offered a suggestion. He said that the *sao* [the fascine bundle] was too long, so long that it was beyond the power of human strength to press it down into position at the bottom of the gap (against the force of the current). The water was rushing through incessantly, so that the cables frequently broke. 'What we ought to do', he said, 'is to separate the *sao* into three sections each twenty paces in length, connected together by ropes. After the first *sao* has been put into place on the bottom, the deposition of the second *sao* can be started, and finally

the third section (will close the gap).' Some of the old-fashioned
workers disagreed with him, and insisted that the suggestion should
not be carried out, saying that a *sao* of only twenty paces would
not stop the water's flow, and having three sections would only
double the expenses to no benefit. (Gao) Chao replied: 'The first
sao will of course not stop the flow of water entirely, but it will
abate its impetuosity by half, so that the second *sao* will need only
half the force to be pushed into position (and held there). After
the second *sao* has been fixed, the flow will again be less, so that
the third (and last) *sao* can be pushed into place from level
ground, and the full force of all available manpower can be applied
to it. The two upper *sao* will soon be silted up with sand and mud
from the turbid river, so that little further work will be required.'
      (Guo) Shenxi did not accept Gao Chao's views, and adhered
to previous practice. At this time Jia Weigong was military
commander of the southern districts. He alone thought that (Gao)
Chao was right, and sent several thousand people with confidential
instructions to points lower down to collect stones and brushwood
(to demonstrate that Guo Shenxi's filling material was simply being
carried away as fast as it was put down). And after the *sao* was put
in position the *sao* itself (began to break up, and fragments of it
were) carried away. And the (Yellow) River continued to flow
through the gap worse than before. So (Guo) Shenxi was recalled
and exiled, and eventually the suggestions of (Gao) Chao were put
into practice, so that Shanghu had peace (from floods).

This gives a revealing insight into discussions which must have gone on for
many centuries among those concerned with water control. Guo Shenxi's
advisers seem not to have been practical men, while Gao Chao was evidently
a man of the people, for in China there was an upsurge of invention and
technical innovation among the masses. This reminds one of the famous
phrase in the *Shen Zi* (The Book of Master Shen) (*c.* second to eighth
centuries AD): 'As for those who protect and manage the dykes and channels
of the nine rivers and the four lakes, they are the same in all ages; they did
not learn their business from Yu the Great, they learnt it from the waters.'
   From the tenth century AD onwards, there was a great succession of
engineers who achieved remarkable results in controlling the Yellow River,
and brought the utmost ingenuity to bear on closing breaches in the dykes.
Gabions and fascine bundles were delivered from boats, boats were filled with
stones and sunk where necessary, and mobile scrape-dredgers were (as we
have seen) constructed. How far river training was included in their art is,
at present, uncertain, but one would guess that their experience in this must

have been considerable. For example, in AD 1015 an imperial edict laid down that in certain sections the depth of the Bian Canal was not to be less than 2.3 metres, but Ma Yuanfang, then Vice-Director of the Court of Imperial Sacrifices, urged that 1.5 metres would be enough as long as the average breadth of 15 metres was adhered to. Besides towpaths, he made jetties, groynes or spurs, and so on, all designed to direct the force of the current away from the concave eroding bend so as to scour the shoals on the convex side. In consequence, banks were saved and the fairway for vessels improved. All this was certainly standard practice during the Ming and Qing.

Sluice-gates, locks and double-slipways

After it was realised that water, even on a large scale, could profitably be made to run along channels between permanent ridges, the need must soon have been felt for some kind of obstruction which was readily moveable at will. Thus did the water-gate come into being. The sluice-gate was derived from irrigation and flood control, and its function when transport was involved led to the lock-gate; the only difference between the two being that the lock-gate was constructed to allow shipping to pass through.

The invention of the lock took place in a number of stages, first by placing widely spaced 'flash-lock' gates (see p. 176) along canals or canalised rivers, then with the appearance of the pound-lock not much longer than the barge or boat which it was intended to raise or lower. This is an arrangement which greatly reduces the time which a vessel must wait while the change in water-level is completed. Finally, improvements to pound-locks such as a mitre type of construction, closing so as to oppose the maximum strength of the current, or the development of ground sluices (passages in the masonry of the lock to admit or remove water without movement of the gates themselves, or of wicket-doors in them), or even of side ponds (reservoirs on both sides of the lock to conserve half the water at each operation) were made.

No mention of gates occurs in the *Historical Records* but they do appear in the History of the Former Han Dynasty. Indeed, in about 36 BC, Shao Xinchen incorporated a number of sluice-gates in his Qianlu reservoir and canals near Nanyang (see p. 197), while a speech made by Jia Rang in 6 BC shows that the use of gates cannot have been a new idea at that time. Jia Rang was submitting a memorial about controlling the Yellow River, and in this he refers to the Bian Canal, where he envisages sluice-gates large and small, set in stone piers, not as a new suggestion but as an improvement on wood and earth ones which had long been used (p. 197). In fact, there had been gates at the entrances of the Bian Canal in the second or first centuries BC, and perhaps also at various points along its course, while there remains always a certain possibility that Shi Lu introduced them on the approaches to the Magic Canal as early as the third century BC.

All this makes it clear that there had been flash-lock gates along the Bian Canal at least by the end of the first century BC, that Wang Jing restored them in the first century AD, and that almost seven centuries later the Ordinances of the Tang department of Waterways contains many articles concerned with sluices and flash-lock gates; '. . . where there are water-mills, the mill-owners should be made to construct gates to regulate the flow of water, and allow free passage along the waterway (for traffic).'

This was in AD 737, but it presaged exactly the arrangement in the sixteenth and seventeenth centuries AD in Europe, where mill weirs were still accompanied by stanches or flash-lock gates to allow vessels to pass up or down stream. The 'flash' of water released by opening the gate was often essential for taking a barge over the shallows lower down, but after opening there might be a wait of a couple of hours for the 'abatement of the fall' before a barge could go up again. There are also plenty of poetical references to flash-lock gates in later Chinese literature. Thus in about AD 1200 Zhang Zi has a poem written beside a river or canal near a temple at Lingyin Shan; when the gate was opened to let a sampan through, the water came pouring down with a thunderous noise.

In the seventeenth century AD foreign travellers began to take notice of flash-lock gates in use on the Grand Canal and other Chinese waterways. They saw vessels hauled through upstream, generally with manned capstans and tow ropes, against currents of as much as nine or ten knots, and in the reverse direction saw them being allowed to 'shoot the rapids'. Fig. 489 reproduces a drawing which accompanied an account of the Macartney embassy, the first official British delegation to the Chinese capital at Beijing, made in AD 1793.

There is no doubt that throughout Chinese history the most typical form of sluice- and lock-gate was what is called the stop-log gate. Fig. 490 shows a small field example, and in Fig. 491 we see a larger version used for flash-locks in canals and canalised rivers. Two vertical grooves cut in wood or stone face each other across the waterway (Fig. 492), and in them slide a series of logs or baulks of timber let down or withdrawn as desired by ropes attached to each end. Windlasses or pulleys in wood or stone mountings acted like cranes on each bank, helping to remove the gate planks. This system was sometimes improved by fastening all the baulks together to form a continuous surface and then raising or lowering the structure in its grooves by counterweights on the ends of cables (Fig. 493). Evidence in the *Glowing Mirror of the Flowing Waters* of AD 1725 indicates that grooved gates were nearly always used, but apparently gates that swung sideways, such as those familiar to Europeans, were not employed in China until modern times. Today, however, the counterweighted steel shutter gates rising and falling after the Chinese style are used in waterways all over the world.

Fig. 489. Flash-lock gates on the Grand Canal in AD 1793 as seen by the Macartney embassy. (Picture from Staunton (1797).)

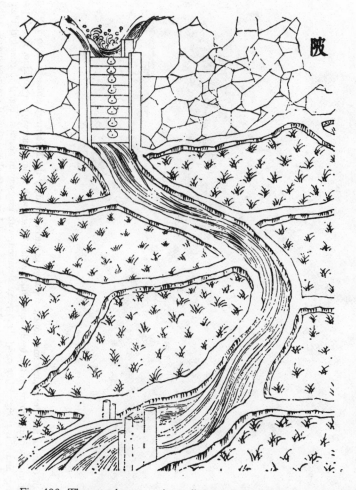

Fig. 490. The stop-log gate. A small example in a dyke, hence the caption (top right) *bei*, forming part of a minor irrigation system. From the *Exploitation of the Works of Nature* of AD 1637, but only in versions after 1726.

The simplest name for water-gate in Chinese literature is just that, *shui men* [*shui mên*] (水門). As time went on, other names were introduced; for instance, 'dipping gates' (*dou men* [*tou mên*] 斗門), though this became obsolete by the fourteenth century AD. We find *zha* [*cha*] (閘) for a lock-gate, and *ban zha* [*pan cha*] (板閘) for a lock-gate made of boards, i.e. a stop-log gate. The term *xuan men* [*hsüan mên*] (懸門), or 'hanging gate', seems

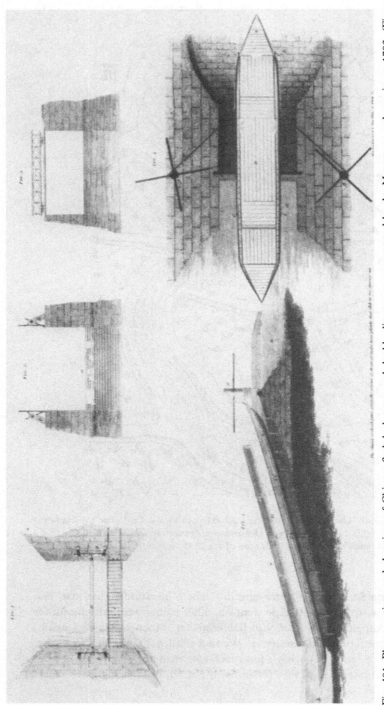

Fig. 491. Plans, sections and elevations of Chinese flash-lock gates and double-slipways as encountered by the Macartney embassy in AD 1793. (Fig. 1 [*upper left*] Plan of flash-lock (stop-log) gate, looking down on the baulks and roller winch hoists, with rolling gangways in position. Fig. 2 [*centre, top*] Section of flash-locks, showing top-logs and hoists. Fig. 3 [*top right*] Transverse elevation, with gangway in position. Fig. 4 [*bottom left*] Plan of double-slipway, showing stone revetments of the inclines, and two capstans. Fig. 3 [*bottom right*] Longitudinal elevation of double-slipway, with boat ascending to the higher level, one capstan shown.) (Picture from Staunton (1797).)

Fig. 492. The Qingfeng flash-lock gate at Gaobeidian near Beijing, showing the grooves in the stonework for the stop-log baulks. (Photo. from Wang Pi-Wên (1935).)

Fig. 493. One of a pair of stone crane arms with roller hoists for lifting and lowering the stop-log baulks, at the Qingfeng flash-lock gate. The canal bed in the background is almost dry. (Photo. from Wang Pi-Wên (1935).)

first to appear in AD 984; this is significant as indicating the permanent installation of windlasses. All these (and some others) could be used indifferently for sluices and flash-lock gates. In the Song period, about the beginning of the eleventh century AD, *cha* [*zha*] or *cha* [*chha*] (牐) both make their appearance, just about the time of the invention of the pound-lock, but this may be a coincidence.

The invention of the pound-lock is a question of substantial importance in the history of civil engineering. Industry and commerce in paleotechnic Europe – the time of crafts and hand tools – were greatly affected by the simple and convenient device of arranging gates so close together as to admit only one or two barges, thus allowing changes of water-level to take place in the shortest possible time, with losses of water from the upper levels accordingly reduced. This key invention, so simple and yet so vital for intensified water-borne traffic in hilly country, invites a comparative study of its appearance in different civilisations. The fact that foreign travellers on the Grand Canal from the seventeenth century AD onwards spoke only of flash-lock gates and double-slipways has seemed to some to be sufficient evidence that pound-locks were never developed in China. But this assumption would fall into the trap of supposing that inventions once made were necessarily utilised in Chinese culture, whether or not the need for them continued. In fact, it is possible to show that the pound-lock originated in China earlier than anywhere else, but was little used in later times because need for it ceased as conditions changed. Dr Needham believes he is able to suggest how this happened.

The oldest example of a pound-lock in China dates from the beginning of the Song dynasty, and is connected with the name of Qiao Weiyao, Assistant Commissioner of Transport for Huainan in AD 983, a man who deserves to be remembered. He was concerned with the barge traffic problem at the northern end of the Shanyang Yundao section of the Grand Canal between the Yangtze and Huaiyin, and it is interesting to find that his invention arose from a social cause, his exasperation with the thefts of tax-grain made possible by the high casualty rate of ships crossing the double-slipways. The *History of the Song Dynasty* says that, in AD 984,

> Qiao Weiyao also built five double-slipways (lit. dams, *yen*) between Anbei and Huaishi (or, the quays on the Huai waterfront). Each of these had ten lanes for the barges to go up and down. Their cargoes of imperial tax-grain were heavy, and as they were passing over they often came to grief and were damaged or wrecked, with loss of grain and embezzlement by a cabal of the workers in league with local bandits hidden near by.
> Qiao Weiyao therefore first ordered the construction of two gates

at the third dam along the West River (near Huaiyin). The distance between the two gates was rather more than 50 paces (76 metres), and the whole space was covered over with a great roof like a shed. The gates were 'hanging gates'; (when they were closed) the water accumulated like a tide until the required level was reached, and when the time came it was allowed to flow out.

He also built a horizontal bridge between the banks, and added dykes of earth with stone facings to protect their (or its) foundations. After this was done (to all the double-slipways) the previous corruption was completely eliminated, and the passage of the boats went on without the slightest impediment.

Such were the first pound-locks in the history of any culture. Large enough to take several vessels at a time, they must have been somewhat similar to those depicted in a well-known illustration by Vittorio Zonca in AD 1607 showing a lock basin on the canal which connected Padua with the River Brenta.

Contrary to an impression that has sometimes prevailed, Qiao Weiyao's pioneering led to a great wave of interest in the new technique. This we know from an informative passage in Shen Gua's *Dream Pool Essays* finished in AD 1086. In this, he explains how the use of pound-lock gates increased the tonnage that each barge could carry from some 21 tonnes to 28.5 and then, in his own day, to up to 50 tonnes. He refers also to an inscription by Hu Su dated AD 1027 saying that the work referred to was probably accomplished at Zhenzhou in 1025.

The next evidence is equally demonstrative and contains a little more detail. It is by the Japanese monk Jōjin who journeyed northward on the Bian section of the Grand Canal in August AD 1072. He says,

> Weather fine. At the *mao* double hour (about 5 a.m.) our boat cast off. By the *wu* double hour (11 a.m.) we got to Yanguan xian [Yen-kuan hsian], arriving at the Chang'an double-slipway. About the *wei* double hour (1 p.m.) the magistrate came, and we took tea at the Chang'an rest house. About the *shen* double hour (3 p.m.) two of the lock gates were opened (in succession), in order to let the boat through. When it had passed through, the stop-logs were dragged back so as to close (the middle gate), and then the stop-logs of the third lock gate (were lifted out) to open it, and the boat was let through. The surface of the succeeding part of the canal was a little more than 1.5 metres lower (than the upper part). After each gate was opened, the (water from the) upper section fell and the water level became equal, whereupon the boat proceeded through.

Here then we have a description of a two-stage pound-lock with three confining gates, the gates being set so close together as to take Jōjin's barque alone, or with a few others. In the entries in his diary for the following month he describes three pound-locks of the more normal two-gate type. Finally, on his return journey in April 1073, Jōjin notes that

> A letter came from the Transport Bureau to say that the
> gate-keepers must wait for a hundred boats to collect, but if in
> three days' time that number had not been reached, then they
> might open the gates. For the needs of irrigation were not to be
> neglected.

Jōjin also describes a clear case of a pound-lock on a canalised river, the Yue. Statements about pound-locks can also be recognised in Chinese testimonies over the next three or four centuries, even when these are less clear than those of Jōjin and even Shen Gua.

In the light of the facts so far described, it is clear that no European work can compete with the Chinese for priority. However, it is possible that the sluice-gate may have originated in the Fertile Crescent of Mesopotamia in the first millennium BC, though matters are not certain, but there seems little reason to doubt its existence in the impressive sweet water canal system that Sennacherib gave to Nineveh in 690 BC, though this is not completely proved. However, even stronger evidence going back further, to the second millennium BC, is provided by the Phoenician harbour works which still exist at Sidon, where rock-cut grooves (of which only one now remains) indicate the former presence of sluice-gates. As far as Egypt is concerned, there may have been a single flood-gate in a canal which connected the Nile with the Red Sea, begun 610 to 595 BC but not completed until 280 BC.

After this there is complete silence in the Western world concerning sluices or locks until the Dutch archives begin to deliver information from the eleventh century AD onwards. Naturally, floodgates (stanches) or flash-locks appear first, stanches in Holland in AD 1065 and for the River Scarpe in Flanders in 1116, but it is not certain that these were used for aiding navigation, though quite probably they did, like the flash-lock gates in Italy on the River Mincio near Mantua by 1198. Floodgates, or tide sluices, permitting vessels to pass between canals and tidal rivers, developed early in the Low Countries, as at Damme near Bruges in AD 1168, on the Zuider Zee soon after 1285, and the well known 'magnam slusam' at Nieuwpoort, active in 1184. In general it may be said that flash-lock gates were common on European canals and canalised rivers by the end of the thirteenth century AD.

Next, it is possible to pinpoint the development of pound-locks with some accuracy. They first arose in places where there were differences of water-level only, that is on the highly tidal shores and estuaries of the North Sea. The first certain date is AD 1373, when a pound-lock was built at Vreeswijk in Holland at a point where a canal from Utrecht joined the River Lek. A similar basin at Spaarndam (replacing the tide sluice) may possibly have existed from 1315, but was definitely there in 1375. Both these were large basins accommodating twenty or thirty ships. The earliest European small basin, and therefore a pound-lock of truest type, dates from 1396 at Damme near Bruges, where again it replaced tidal sluices; its length was 30 metres.

It was at just about this time that the first successful attempt was made in Europe to overcome variations in the ground level, that is to say by the building of a true summit canal. This, again, was in the Low Countries. The Stecknitz Canal, completed in AD 1398, bestrode a watershed difference of some 17 metres with the aid of two considerable pound-locks. Only after this came the upsurge of Italian civil engineering. A great builder then of pound-locks was Bertola da Novate, who constructed twenty-three on part of the Milan canal system between 1452 and 1459. He suggested gates moving horizontally, not vertically, and this was taken up by Leonardo da Vinci, who certainly invented the mitre gate and built several examples of it before 1497. Leonardo also included a wicket door for admitting water with a valve balanced eccentrically like a Chinese rudder (see this abridgement, volume 3, p. 233). It was Leonardo who, between 1497 and 1503, also suggested installing a series of locks for carrying traffic over a watershed summit canal.

In the light of all this, it is paradoxical that post-Renaissance travellers on the great Chinese canals should have found only flash-lock gates and double-slipways. The explanation seems to be implied in the quotation from Shen Gua mentioned on p. 247, when he discussed the tonnages of tax-grain boats using the Bian Canal, and how they increased rapidly when pound-locks were substituted for slipways. Pound-locks were, in fact, essentially the response to heavily laden vessels on the canals, and if a time came when this stimulus should be withdrawn they might well not be renewed.

This is exactly what happened. When in the late thirteenth century AD the Yuan government fixed its capital at Beijing, the canal system could not at first, and indeed, throughout that dynasty never wholly, carry the weight of the northbound traffic, so that a large part of it was shipped by sea (see p. 222). The annual sea shipments often equalled the inland totals and quite often exceeded them. After AD 1450, Ming rule let drop the naval might that the Southern Song and Yuan dynasties had laboriously built up, but this was because during the previous three centuries the need for really heavy craft on the Grand Canal, like those which the Northern Song had employed with such striking results, had been in abeyance. As a result, it had become customary to use a multitude of smaller vessels, and as this tradition established itself, the pound-locks fell into decay one by one and were not replaced.

The decline and fall of the pound-lock in China notwithstanding, what influences could Chinese hydraulic engineering have exerted on Europe, given the datings just established? It is always conceivable that the twelfth century AD Netherlanders and Italians knew by Crusader gossip of the Chinese flash-lock gates of Wang Jing, Li Po and their innumerable colleagues, going back at least to the first century AD, but the device was of such natural simplicity that they could have invented it themselves. Perhaps, however, the pound-lock is another matter.

It may not be a mere coincidence that the beginnings of pound-locks and summit canals in the West occur in the fourteenth century AD, just after the 'Pax Mongolica' permitted the free travel and intercourse of merchants typified by Marco Polo. Moreover, many Chinese were in the first wave of the Mongolian conquest of Iran and Iraq – a Chinese general, Guo Kan, was first governor of Baghdad after its capture in AD 1258. As the Mongols had a habit of destroying irrigation and water conservancy works of all kinds to annoy their more agricultural enemies, it would have been most natural for them to turn to their Chinese technical colleagues for rehabilitation as soon as government exploitation was ready to replace military operations.

At some quite early time, it must have been realised that if the ramp of a spillway were made to slope at a reasonably gentle gradient, it would be possible to drag canal boats up and over it to a higher level, wastage of water at the same time being prevented. Thus arose the double-slipway, a pair of inclined stonework aprons over which boats were hauled, generally in China with the use of capstans, from a waterway at one level to that at another. The thing was also inspired, no doubt, by the primitive practice of ancient peoples of carrying boats to avoid rapids on rivers, or to connect two arms across an isthmus. The commonest Chinese term for the double-slipway was *dai* [*tai*], but this seems also to have sometimes been used for dams and dykes, so that its interpretation may raise difficulties. However, when in AD 384 the philosophical Jin general Xie Xuan built seven *dai* on the Bian Canal near modern Tongshan, he did it 'to facilitate transport', so that slipways are certainly meant. Such slipways were still in widespread use when the European travellers began to write their narratives of Chinese journeys. In AD 1696, Lecomte gave a fine description:

> I have observed in some Places in China, where the waters of two
> Canals or Channels have no Communication together; yet for all
> that they make the Boats to pass from the one to the other,
> notwithstanding the Level may be different above fifteen Foot
> (4.5 metres): And this is the way they go to work. At the end of

the Canal they have built a double Glacis, or sloping Bank of
Freestone, which uniting at the point, extends itself on both sides
down to the Surface of the Water. When the Bark is in the lower
Channel they hoist it up by the help of several Capstanes to the
plane of the first Glacis, so far, till being raised to the Point, it
falls back again by its own Weight along the second Glacis, into
the Water of the upper Channel, where it skuds away during a
pretty while, like an arrow out of a Bow; and they make it descend
after the same manner proportionably. I cannot imagine how these
Barks, being commonly very long and heavy laden, escape being
split in the middle, when they are poised in the Air on this Acute
Angle; for, considering that length, the Lever must needs make a
strange effect upon it; yet I do not hear of any ill Accident happen
thereupon.

COMPARISONS AND CONCLUSIONS

The comparisons of the previous pages invite a more extended survey of the
achievements of Chinese hydraulic engineering in relation to that of other
times and places – Mesopotamia and Egypt, Greece and Rome, the Renaissance, and so on – for which this book is hardly the place. But some
examination by scholars suggests that, when allowance is made for the available technical means, only the systems of Babylonia and ancient Egypt, and
in later times India and Sri Lanka, can be compared with what was accomplished in China.

Irrigation engineers distinguish between perennial and inundatory canals,
the first drawing water for the fields at all times of the year, the second filled
only during the flood season. This differentiates between the systems of
ancient Mesopotamia and ancient Egypt. In the former, the plain of the
Tigris and Euphrates rivers was watered at all seasons, but in Egypt there
was a vast chain of retention basins where silt was deposited, and these were
(and are) filled for only forty-five days during the year. This can be done
because the Nile – the 'most gentlemanly of all rivers' – gives ample warning
of its rise and fall, makes no abrupt changes, and has enough silt to enrich
the land annually but not enough to choke the channels. It is also free from
salt, and it flows between sandstone and limestone hills, which furnish
unlimited building stone. The Tigris and the Euphrates possess few of these
qualities, and the Yellow River almost none.

Broadly speaking, the hydraulic works of the great civilisations of South
and East Asia combined in various proportions the Egyptian and Babylonian
patterns to form more flexible systems. In India, the fate of water conservancy projects was far less happy than in China, owing partly to the absence
of continuing political and linguistic unity and centralisation, and partly to

the recurring disruptions of the country by foreign invasions. But there can be no doubt that canal building is ancient in India, and that a waterway made there in the fifth century BC may well not have been the first. Yet it was never in India that the fusion of the Egyptian and Babylonian patterns achieved its most complete and subtlest form.

This occurred in Sri Lanka, the work of two cultures, the Sinhalese and the Tamil, but especially the former. This was because of both the meteorological and the geographical nature of the island. The central mountains of Ruhunu and Maya Ratta are surrounded almost on three sides by 'dry jungle', watered only seasonally by the south-west and north-east monsoons, respectively. But the mountains themselves, together with the tract of country constituting the south-west quarter of the island, receive a rich rainfall as well as some perennial rain. For this reason, both Mesopotamian and Egyptian methods were required.

It seems that what occurred was that the farmers made numerous small tanks in the hills and foothills near their fields or terraces to catch the run-off water, which they baled out at leisure. Then numbers of small dams forming small reservoirs were built, often in series, on the upper reaches of the greater rivers, thus retaining the annual inundation, and discharging it as desired by small canals along the valley sides. As time went on, larger reservoirs were built, submerging or rendering unnecessary the smaller ones. The next step was revolutionary: a weir was built much higher up the main river to form the headwork of a long contour derivation-canal, which thus brought perennial water to join the annual monsoon supplies in the great reservoir. This canal, and others like it, often crossed one or more watersheds on their way. They generally had dykes only on one side, sometimes spreading out into small lakes as they went, though in some places a double embankment was needed. Smaller tributaries and gullies were crossed by spillway dams with wing walls, sufficient to take care of the greatest freshets of water, but also arranged to deliver a constant supply from the canal in dry periods, thus converting a fitful tributary into a perennial stream. This saved the labour of building purely artificial distribution canals. Elsewhere, the contour derivation-canals crossed apparently flat country for many kilometres. All these works can be traced today, and some still function.

Sinhalese hydraulic engineering was not born at quite the same time as Sinhalese Buddhism, but very nearly so. For the mission of the apostle Mahinda – the younger brother of King Asoka, the last emperor of the Maurya dynasty in India – began in about 250 BC, just about a decade before the Tissa-wewa, one of the large man-made lakes, was constructed. This, of course, was contemporary with the State and Empire of the Qin and the Zhengguo and Guanxian irrigation systems as well as the Magic Canal.

The vast majority of the other waterworks in Sri Lanka were constructed over the next 1000 years, and little was done after the twelfth century AD.

The Sinhalese engineers also built trunk canals, which do not end in reservoirs directly, but convey water from a high division-head along a lateral course above a river, providing irrigation water on the way. By the first century AD, they understood the principle of the oblique weir, designed to guard against shocks from the water which might dislodge the masonry. Moreover, they faced the inside surfaces of their reservoir embankments with ripple bands (stone revetments to prevent wave-erosion). But perhaps their most striking invention was the intake or valve towers fitted to their reservoirs, perhaps from the second century BC, certainly from the second century AD. These caused the water to spill down their walls and leave the tank by double horizontal sluice-tunnels or culverts at the base of the embankment. In this way, water free of silt or scum was provided, with the pressure reduced so as to render the outflow controllable.

Such were the heights of South Asian hydraulic engineering in ancient and medieval times. When we turn to the East Asian theatre and look back upon all the Chinese achievements, we are struck by a marked difference in emphasis. Irrigation, though important, is no longer supreme, and has to share prominence with an unceasing struggle for river-control and a constant preoccupation with inland water transport.

A couple of questions remain. If one asks where the works first arose, one has to visualise initially, in early Zhou times (1030 to 722 BC), a wide area from the lower Yangtze basin northward to the lower Yellow River. Then in the Warring States time (221 to 207 BC) the addition of the north-west (Guanzhong, the Wei valley) and western (Sichuan) areas. Already in the Qin dynasty (third century BC), all parts of China, except perhaps Fujian and Yunnan, were open to operations. If one asks what the earliest works were, it would seem that collecting run-off water from hill valleys in tanks by dams, with small derivative canals, came first, quickly followed by river-controlling dykes; then, shortly after, the cross-country navigation canal, perhaps even preceding the river-derived lateral irrigation system. It is quite clear therefore that the Chinese achievements, though conditioned by physical geography and social features, were deeply original, 'a symphony on benefit of water by a different composer'.

If, lastly, one asks how the Chinese picture fits in to the framework of the annual–perennial concept, the answer is not very easy. Nowhere in China were two highly contrasting zones in that proximity which we find in Sri Lanka. With a climate essentially monsoon in character, all China's rivers had a marked rise and fall, for instance the 30-metre excursion of the Yangtze at Chongqing. They were therefore annual in this sense, but perennial in that the low-water levels were themselves formidable; the Tibetan snow fields saw to that. In the great river valleys, there was always the tendency to lead off water for irrigation when the annual flood crest was passing, though this was opposed by many engineers, who wanted the strong current to scour the river bed. As for the most interesting of Chinese systems, that of Guanxian (p. 202), this partook of both types. It was not purely based on inundation; there was always a small flow in the Neijiang River, even during the dry season (except for the time when the canal was closed for scouring the river bed). It was not strictly perennial either, since the difference between flood and dry season conditions is so large. In a word, all possible variations could be found in China, and it was perhaps for this very reason that the invention and widespread use of sluices and lock-gates took place there so early.

In effect, the story of the hydraulic works of China is nothing short of an epic. If the nature of the climate and the soil dictated them, if the form which Chinese society inevitably took necessitated and fostered them, they were nevertheless not achieved without the toil, more often willing than otherwise, of unnumbered millions of men and women, nor yet without the devotion, skill and ingenuity of successive generations of civil engineers worthy to be compared with those of any other people. It was Sima Qian some twenty centuries ago who, in his *Historical Records*, wrote:

> In the south I have climbed upon Mount Lu and seen the Nine
> Rivers with the courses which Yu the Great gave them; I have
> visited the Guiji Mountain and been to Taihuang; I have sat on
> the Gusu (terrace) and contemplated the Five Lakes. In the east
> I have viewed from Mount Dapei the place where the Luo River
> joins the Huang He. I have sailed upon the Yellow River itself,
> and travelled on the canals which join the rivers Huai, Si, Ji, Ta
> and Luo. In the west I have looked towards the mountains whence
> the Min River flows forth and seen the 'Separated Hill' (at
> Guanxian) in the country of Shu (Sichuan). In the north I have
> been beyond the Longmen gorges (of the Yellow River) even as far
> as Shuofang.
> And I say again, inconceivably great are the benefits and
> destruction which water can produce.
> And in my time I myself, when in the imperial suite, carried
> brushwood bundles to help to close the breach at Xuanfang; I
> shared in the sorrow expressed by the emperor's ode that he made
> at Huzi and now at last I have written this treatise on the (great)
> River and the Canals.

6

Hydraulic engineering (II), water-raising machinery and the use of water as a power source

We now turn to consider machinery used for raising water – an operation of immense importance in any civilisation based on irrigated agriculture – then the application of water as a power source for driving all kinds of machinery and, finally, the use of rotary power to propel ships through water.

WATER-RAISING MACHINERY

As far as water-raising machinery is concerned, we shall describe in due order (a) the counterweighted bailing bucket, (b) the well-windlass, (c) the scoop-wheel, (d) the square-pallet chain-pump, (e) the vertical pot chain-pump, and (f) the peripheral or rim pot wheel. This list by no means exhausts the variety of water-raising devices known and used in ancient and medieval times, or traditionally still in employment, but it meets the needs of the Chinese, and reference to other types of water-raising devices will be made incidentally as we go along.

The counterbalanced bailing bucket – the 'swape' or 'shādūf'

The oldest and simplest mechanism which lightened the human labour of dipping, carrying and emptying buckets was the counterbalanced bailing bucket, usually known as the swape, sometimes as the well sweep, and often by its Arabic name of *shādūf*. This was familiar from early times in Babylonia and ancient Egypt, where it has continued in operation till the present day, and makes use only of the lever, involving no rotary motion. A long pole is suspended or supported at or near its centre, like a balance-beam, and while one end is weighted with a stone, the other carries the bucket at the end of a rope or a piece of bamboo. In the Old World its distribution became almost universal. The earliest illustrations we have of it in China are those of the mid-second century AD in the tomb-reliefs of Xiaotang Shan and Wu Liang, often reproduced (cf. Fig. 494). We see it again on the walls of the Qianfodong cave-temples and in the *Pictures of Tilling and Weaving* of AD

Fig. 494. The swape, or counterbalanced bucket lever, for drawing water from wells; a Han representation from the Xiaotang Shan tomb-reliefs, c. AD 125. Such pictures generally show it, as here, in association with a butcher's shop and kitchen of a great house. (Photo. from Chavannes (1893).)

1210, the *Tian Gong Kai Wu* (Exploitation of the Works of Nature) and other books (Fig. 495). But the earliest mention of it in Chinese literature was in the fourth century BC in an interesting passage in *Zhuang Zi* (The Book of Master Zhuang), which has already been quoted (volume 1 of this abridgement, chapter 8) with regard to the paradoxical anti-technology complex of the Daoists. A farmer declines to use the device when Zigong, a disciple of Confucius, suggests that he should do so:

> Zigong had been wandering in the south in Chu, and was
> returning to Jin. As he passed a place north of the Han (river),
> he saw an old man working in a garden. Having dug his channels,
> he kept on going down into a well, and returning with water in a
> large jar. This caused him much expenditure of strength for very
> small results. Zigong said to him, 'There is a contrivance by means
> of which a hundred plots of ground may be irrigated in one day.
> Little effort will thus accomplish much. Would you, Sir, not like to
> try it?' The farmer looked up at him and said, 'How does it work?'
> Zigong said, 'It is a lever made of wood, heavy behind and light in
> front. It raises water quickly so that it comes flowing into the
> ditch, gurgling in a steady foaming stream. Its name is the swape.'
> The farmer's face suddenly changed and he laughed, 'I have heard
> from my master', he said, 'that those who have cunning devices
> use cunning in their affairs, and that those who use cunning in
> their affairs have cunning hearts. Such cunning means the loss of
> pure simplicity. Such a loss leads to restlessness of the spirit, and
> with such men the Dao will not dwell. I knew all about (the
> swape), but I would be ashamed to use it.'

This might be taken as evidence that the swape reached China in about the fifth century BC.

Batteries of swapes raising water in successive levels are often seen in Babylonian and ancient Egyptian representations, described in Arabic manuscripts, and are even photographed by contemporary travellers. A development of them was to elongate the bucket's spout into a 'flume', often made from a hollowed out trunk of a palm tree, this being linked parallel with a counterweighted beam above, and so arranged that it automatically empties itself into the receiving channel on the upward motion. This is the Bengali *dūn*. In India, the operation of the device is assisted by a moving counterpoise, that is by men who walk back and forth along the upper beam. Finally, the bucket, flume and counterpoise were combined in a single unit, or the place of the counterpoise was taken by manhandled ropes or gearing.

These machines have sometimes been regarded as Muslim inventions because many of them occur in Arabic texts, such as the famous book by

Fig. 495. A swape, or well-sweep, from *The Exploitation of the Works of Nature* of AD 1637. The counterweight can be seen on the left, near the tree.

al-Jazarī of AD 1206. But in view of their wide distribution and frequency in India, they may more probably have originated there. Significantly, they occur in various forms in Indo-China, but not further north. How far such machines were ever common in medieval Islam we do not know. At all events, the flume-beamed swape interested European engineers in the six-teenth century AD and was introduced to China by the Jesuits, though it was probably never employed there.

The swape is not quite such a jejune machine as it seems at first sight, for its exact converse, the movement of a weight by an alternating water counter-balance, has interesting relations with the development of water-wheels. It was also used for raising beacon fires on high. And if certain important types of military catapults were essentially swapes, so also was that great herald of industrial technology, the Newcomen pumping steam engine.

The well-windlass

Rotary motion came in with the pulley or drum set at the mouth of a well. At first, the rope was simply hauled over it and gathered, then the bucket was counterweighted, and finally the drum was turned by a crank. Han tomb-models show the first of these stages (Fig. 496); in this example, one can see the pulley in its bearings at the top under a small tiled roof; the frame is ornamented with dragons. Fig. 497 shows an interesting, though rather bad, drawing of the drum and crank from the seventeenth century AD *Exploitation of the Works of Nature*, but it is not as bad as it looks, for it corresponds remarkably closely with a photograph of a well-digger's wind-lass taken by R. P. Hommel. However, Hommel was not alone in finding that in China the drum generally carried two ropes, so wound that the counterweight or the unfilled bucket descended as the filled bucket came up. Possibly this gave rise to what has been called the 'Chinese windlass', which gains mechanical advantage (see this abridgement, volume 4, pp. 59 and 61).

Other evidence which makes it clear that these simple machines were common in the Han may be found in texts such as the *Book of (the Prince) of Huainan*, where it is recommended not to plant catalpas near wells, for the roots or branches of these large trees of rounded shape will impede the movement of the rope and buckets. It is also probable that large capstans worked by animals were used for hauling up the long bamboo buckets of brine in the salt-well boreholes at this period. Certainly, salt derricks with pulleys at the top are seen in several Han representations.

The scoop-wheel

The next simplest machine is a hand operated paddle-wheel sweeping up water into a flume or channel. Fig. 498 shows the illustration of it in the

Fig. 496. Han pottery jar
representing a well-head with
pulley and bucket. (Photo. from
Laufer (1909).)

Fig. 497. Well-windlass with crank, from *The Exploitation of the Works of Nature* of AD 1637. A plan of irrigated plots superimposes itself on the left of the perspective drawing.

刮車

Fig. 498. Scoop-wheel, or hand operated paddle-wheel, sweeping water up into a channel, or flume, from the *Complete Investigation of the Works and Days* of AD 1742. This was effective only for small lifts.

Shou Shi Tong Kao (Complete Investigation of the Works and Days) of AD 1742, where, as in the *Nong Shu* (Treatise on Agriculture) of AD 1313, it is called *gua che* [*kua chhê*] (刮車). It could be effective only for short lifts. Its simplicity may, however, be deceptive, and it would be unsafe to assume that it was invented before the water-wheel or the paddle-wheel. Though like a paddle-wheel in action, it is not to cause motion over water, but of transmitting motion to water. A version of the scoop-wheel in which the operator treads on its outer rim became particularly popular in Japan. Indeed, it is said to have been invented by two townsmen of Osaka some time between AD 1661 and 1672, and Dr Needham has found nothing similar in any Chinese book, though the treadmill scoop-wheel is in fact widely used (or was until recently) in the salt evaporation areas of East China. It seems also to have become popular in late seventeenth and eighteenth century Korea. The principle was also widely used in Holland and in the English fen country from the sixteenth century onward, where such scoop-wheels were mounted at the bases of windmills. Dr Needham suspects that the paddle-wheel and scoop-wheel came into use some time between the Han and the Tang (AD 618 to 907). Possibly it was introduced to Europe as well as Japan and Korea from China.

The square-pallet chain-pump and the rag-and-chain pump

We come now to the most characteristic of Chinese water-raising machines, the square-pallet chain-pump. This device, as it was found in 1943, is seen in Fig. 499; it consists of an endless chain carrying a succession of flat plates or pallets which draw the water along as they pass upward through a trough or flume, discharging into an irrigation canal or field at the top. It is known as the *fan che* [*fan chhê*] (翻車) ('turnover wheels'), *shui che* [*shui chhê*] (水車) ('water machine'), or, colloquially, *long gu che* [*lung ku chhê*] (龍骨車) ('dragon backbone machine').

Depending on the length of the trough, the square-pallet chain-pump will lift water up to some 5.5 metres, the limit being set by the leakage and the properties of the woodwork. The best inclination is 24°, but in practice it is usually somewhat less. The chain is turned by one of four methods: the human hand or foot, animal-power or water-power. Of these, the oldest is probably the human foot, since the upper sprocket wheel can carry on its axle so conveniently the radial treadle pads; this is seen in all the agricultural treatises since the *Pictures of Tilling and Weaving* of AD 1210. A nearer view of the sprocket-wheel, pallets, chain and treadle pads is shown in Fig. 500, while the best Chinese illustration is probably that given in *The Exploitation of the Works of Nature* (Fig. 501). The same seventeenth century AD work also shows a smaller version of the machine operated by hand using a connecting rod and eccentric lug or crank on the upper sprocket-wheel (Fig.

Fig. 499. Typical Chinese square-pallet chain-pump, which raises water in a flume, the pallet chain passing over a sprocket wheel powered by two or more men stepping on radial treadles. (Photo. Joseph Needham, 1943, taken between Yongchang and Yongchuan, Sichuan.)

502). As far as animal-power was concerned, the animal was attached to a bar or whippletree and a cogged drive wheel, which engaged at right-angles with a gear wheel on the sprocket-wheel axle (Fig. 503); this appears first, however, in a painting of about AD 965. The last method, the horizontal water-wheel, was apparently less used than the others; it again used similar gearing to that adopted for animal-power (Fig. 504). This was first illustrated in AD 1313.

References to a specific machine in any general literature are always rare, but as the square-pallet chain-pump was probably the favourite water raiser in China, we are not so badly off. A text from the fourth century BC gives us a hint, which suggests that this date might be near the beginning of the story. In his famous argument with Mencius about human nature, Gao Zi, as reported in the *The Book of Master Meng (Mencius)*, says:

> Now if you beat water and cause it to leap up, you can make it go higher than your forehead. If you stir it and cause it to move, you can force it high up a mountain side. But are such movements according to the (intrinsic) nature of the water? It is the strength applied (surely), which brings about these effects. When men are

Fig. 500. Detailed view of the mechanism of the square-pallet chain-pump; an example in Anhui.

made to do what is not good, *their* natures are forced in a similar way.

If this is a reference to water-raising machinery (which it has not tradition-ally been taken to be), then the description fits the chain-pump better than any other device, and certainly much better than the swape, which, however,

Fig. 501. The square-pallet chain-pump, from *The Exploitation of the Works of Nature* of AD 1637.

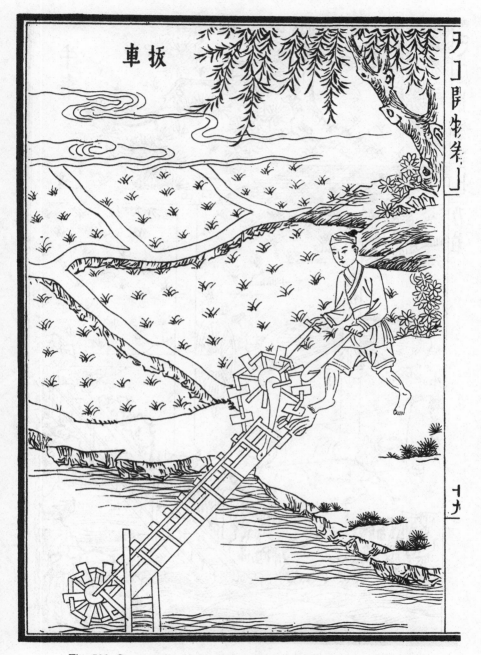

Fig. 502. Square-pallet chain-pump manually operated with cranks and connecting-rods, from *The Exploitation of the Works of Nature* of AD 1637.

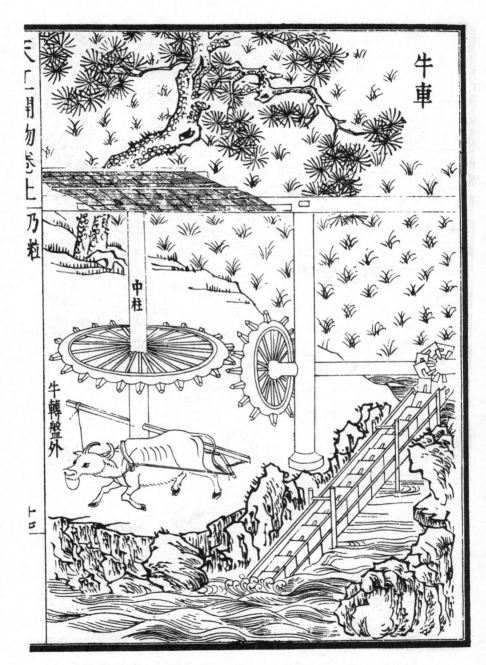

牛車

中柱

牛轉盤外

Fig. 503. Square-pallet chain-pump worked by an ox whim and right-angle gearing, from *The Exploitation of the Works of Nature* of AD 1637. The driving axle is labelled 'central haft', and the legend to the left says that the ox treads a circle wider than the diameter of the horizontal gear wheel.

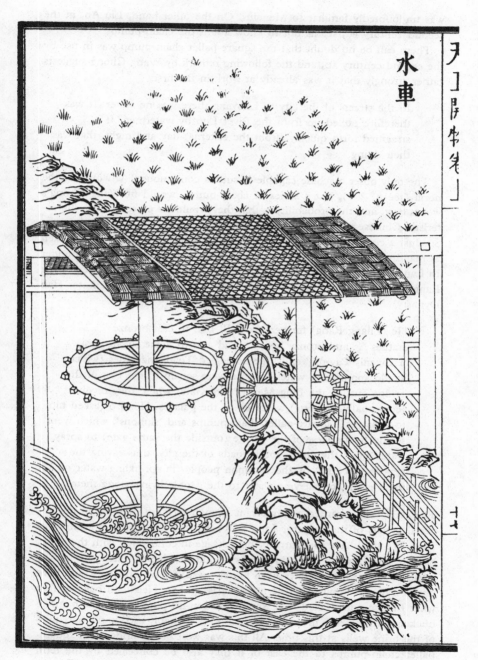

水車

Fig. 504. Square-pallet chain-pump worked by a horizontal water-wheel and right-angle gearing, from *The Exploitation of the Works of Nature* of AD 1637.

was undoubtedly familiar to Mencius. On the other hand, Liu An, in the second century BC, seems not to know about the chain-pump.

There can be no doubt that the square-pallet chain-pump was in use by the second century AD, and the following remark by Wang Chong suggests rather strongly that it was already at work in the first:

> In the streets of the city of Luoyang there was no water. It was
> therefore pulled up from the River Luo by watermen. If it
> streamed forth quickly (from the cisterns) day and night, that was
> their doing.

The only other machine capable of such a continuous flow would be the noria (see p. 277), but a series of chain-pumps would fit Wang Chong's words best, and the probability is that they would have been of the characteristic Chinese square-pallet type. The reference would be from about AD 80. Just a century later, we have the account which has usually been taken as marking the invention itself. It refers to the engineer and master founder Bi Lan, and occurs in the biography of the famous eunuch minister Zhang Rang (*died* AD 189) given in the *Hou Han Shu* (History of the Later Han Dynasty).

> He (Zhang Rang) further asked Bi Lan to cast bronze
> statues ... and bronze bells ... and also to make (lit. to cast)
> 'Heavenly Pay-off' and 'Spread-eagled Toad' machines (which
> would spout forth water). These were set up to the east of the
> bridge outside the Ping Men (Peace Gate) where they revolved
> (continually, sending) water up to the palaces. He also (asked him
> to construct) square-pallet chain-pumps and 'siphons', which were
> set up to the west of the bridge (outside the same gate) to spray
> water along the north–south roads of the city, thus saving the
> expense incurred by the common people (in sprinkling water on
> these roads, or carrying water to the people living along them).

The event was regarded as important enough to merit mention in the imperial annals of the emperor Ling Di's reign for the year in question (AD 186). We shall have to return to this passage in connection with the noria, but the construction of chain-pumps is quite explicitly stated. As for the siphons, the term must have been loosely used, for no siphons would have raised water as implied by the quoted passage. Dr Needham believes that what Bi Lan constructed were simple suction-lift pumps, like the bamboo buckets with valves in their floors, which were probably used in the boreholes of the brine wells at this time. All this was west of the bridge, but square-pallet chain-pumps to the east of it may also be concealed by the term 'spread-eagled toad machine' (*xiama* [*hsia-ma*]), about which more will be said later.

Whatever exactly Bi Lan's water-raising machinery was, we have here, at the end of the second century AD, a valuable account of a rather advanced water supply system for an urban area, and the recent discoveries of so many kinds of stoneware piping and conduits from the Qin and Han give us a clear picture of how it was organised. Such information about Chinese cities has numerous applications, for example in connection with public health and hygiene. There is no doubt that Bi Lan's job was done at Luoyang, and from later sources, such as the *Description of the Buddhist Temples and Monasteries at Luoyang* of AD 530, it is not difficult to identify the site of his works. Deriving from the Gu Shui stream, the Yang Qu, a kind of moat, passed round the city walls to the west and south and was bridged outside the Ping Men gate, which had been built in AD 37. The pumps on its east side were evidently reserved for the palaces within the walls, while those on the west serviced the water-pipes of the city streets. As most of the water seems to have gone round the north of the city to enter the Hong Chi lake, the southern part of the moat was probably very slow flowing – a fact of some importance for our interpretation of Bi Lan's machines, which thus can hardly have been operated by a current of water.

The other classical reference to the square-pallet chain-pump is in the *San Guo Zhi* (History of the Three Kingdoms) of about AD 290, and concerns the famous engineer Ma Jun, who was active at the court of the emperor Ming Di of the Wei:

> Ma Jun of Fufeng was matchless in ingenuity. According to the essay on him by Fu Xuan . . . there was at the capital, within the city, some unused land which could have been made into a park or gardens. But unfortunately no water was available for it. However, Ma Jun constructed square-pallet chain-pumps and had them worked by serving-lads. Whereupon the irrigation water rushed in (at one place) and spouted forth (at another), automatically turning over and over. The skill with which these machines were constructed was a hundred times beyond the ordinary.

This must have been at Luoyang between AD 227 and 239. Later historians, such as the Song author of the *Shi Wu Ji Yuan* (Records of the Origins of Affairs and Things), used to mention Bi Lan and Ma Jun together as the originators of the device. But, apart from the evidence already given, it may be that the chain-pump had already acquired its name of 'dragon backbone machine' in the Han; in which case, some early references to dragon bones, dating from the last decades of the first century BC, may have meant the machine and not the fossil. Such, at any rate, was the idea of some Tang commentators.

In thinking over the first invention of the square-pallet chain-pump, one has to take into account the fact that the expression *fan che* [*fan chhê*] (翻車) did not always have sole reference to this machine. According to that most ancient of Chinese dictionaries, *The Literary Expositor*, using material from the time of the Warring States and compiled in Qin and Han, the 'over-turning device' was another name for a kind of bird-trap called *fu* (罦, 覆) which was more or less synonymous with the *chong* [*chhung*] (罿) or *zhuo* [*chuo*] (罬). On this, Guo Po, at the beginning of the fourth century AD, comments that the *fu che* [*fu chhê*] (覆車) was also called the *fan che* [*fan chhê*] (翻車) and tells us that it had two parallel bars with a net at the far end. Here, the swape or catapult (trebuchet) bird-trap is easily recognised; the hunter, on seeing birds alight, pulls sharply down on a cord attached to the short arm of the lever, thereby lobbing over the long arm and the net attached to it, like the sling of a trebuchet catapult. This device goes back to the time of the *Book of Odes*, of the early Zhou period. Whether it had any connection with the idea of the square-pallet chain-pump is another matter, though it has been suggested (not perhaps altogether implausibly) that Bi Lan was inspired partly by the two bars of the swape bird-net for the sides of his flume, and by its overturning motion for the upper action of his endless chain. Moreover, that curious Zhou idiom for the human jaws (*ya che* [*ya chhê*]) may be related to the action of the chain as it scoops up the water. In any case, the two uses of the term *fan che* deserve to be recorded.

By the Tang and Song the square-pallet chain-pump had become a commonplace, mass-produced by thousands of rustic wheelwrights. In AD 828, its specification was standardised. The *Jiu Tang Shu* (Old History of the Tang Dynasty) written in AD 945 says:

> In the second year of the Taihe reign period, in the second month . . . a standard model of the chain-pump was issued from the palace, and the people of Jingzhao Fu were ordered by the emperor to make a considerable number of the machines, for distribution to the people along the Zheng Bai Canal, for irrigation purposes.

The device was also getting into literature – for example, a rhapsody on it was written by Fan Zhongyan (AD 989 to 1052). By the fourteenth century, it was customary for girls to work the square-pallet chain-pumps, and in about 1145 the painter, poet and official, Lou Shou, devoted one of his poems to it:

The Man of Song, who pulled up the sprouts, we despise;
And Zhuang (Zi) of Meng, who preferred hand-watering with jars.
These are not so good as (that engine, which works)
Like a set of birds, each holding the tail of the next in its mouth.
With this we can change water's flow, and drain a whole lake.
So dance the rice-plants happily in the sky-blue waves
While the farmer sits on his bamboo mat
Enjoying the cool of the evening, and the singing and laughing
Of lads and lasses, under the willows,
Lit by the setting sun.

An early account of the use of animal-power working chain-pumps is given by Lu Yu in describing his journey to Sichuan in AD 1170, though this is preceded by a painting of about 965. And it must be realised that these machines have always been used for raising water in civil engineering as well as for agricultural irrigation. Indeed, they even take their place in Lin Qing's *He Gong Qi Ju Tu Shuo* (Illustrations and Explanations of the Techniques of Water Conservancy and Civil Engineering) of 1836.

The square-pallet chain-pump is probably what is meant in a passage in the travel journal of the Daoist sage Qiu Changchun, written on his way through Turkestan in AD 1221 to visit Chinghiz Khan. Meeting for the first time with cotton, he observed the work of the local farmers:

They irrigate their fields with canals, but the only method
employed by the people of these parts for drawing water (formerly)
was to dip in jars and carry them back. When they saw our
Chinese water-raising machines, they were delighted with them.

Later, in the Yuan period, the square-pallet chain-pump spread even more widely, for in about AD 1362 a Korean official, Paek Munbo, urged that they should be adopted in his own country. In the reign of King Sejong (1419 to 1450) there was much discussion about the advantages of square-pallet chain-pumps and norias, but it is clear that the construction of the former was encouraged by the Korean court from 1400 onwards, while, in the eighteenth century AD, the great progressive Korean scholar Pak Chiwŏn was still advocating the wider use of Chinese water-raising techniques. In the nineteenth century, such pumps spread widely in Indo-China.

The square-pallet chain-pump, however, was destined to achieve a distribution far wider than that of the Chinese culture area. In the UK, visitors to the Palace of Hampton Court near London may see a remarkable square-pallet chain-pump of pure Chinese pattern, which is said to have been installed there for the removal of sewage in AD 1516, but which dates more probably from about 1700. Although the size of the pallets closely resembles that of those used in China, the maker failed to reproduce certain subtleties of the original type, such as greater height than breadth of the pallets, and the arrangement that their grain should be at right-angles to the wearing surfaces.

There can be little doubt that this type of chain-pump spread all over the world from China in the seventeenth century AD. Though something very similar is described by Buonaiuto Lorini as early as 1597, it was only just before 1671 that a visitor with one of the Dutch Embassies to China described water being '. . . conducted from low to high Places by means of an Engine made of four square planks holding great store of Water, which with Iron Chains they hale up like Buckets.' The only fault in the observation was that the chains were of wood, not iron. And it appears that, towards the end of the seventeenth century, chain-pumps were in use on British naval vessels, having been adopted from Chinese ships. In the eighteenth century, there are numerous descriptions in engineering literature, and in 1797 van Braam Houckgeest described his introduction of it to the United States '. . . where it has proved of great utility along the river-banks, on account of the ease with which it operated.'

But to return to China, it seems that by the time of the Song (AD 960 to 1129) the square-pallet chain-pump may have given rise in China to the endless conveyor chain of containers for excavating sand or earth. In any case, with the Renaissance, the idea arose and spread in the West; it was even offered to the Chinese by the Jesuits.

Another obvious application of the endless chain was for dredging, as can be seen from numerous European designs from the sixteenth century onwards, but this use, with its need for scooping buckets rather than pallets, derives presumably form the pot chain-pump or *sāqīya* (see below) rather than the dragon backbone machine. Here there was a junction of techniques. If the dredger was, and still is, set up slanting like the latter, it has buckets like the *sāqīya*. The obvious converse intermediate device would have the pallets (or their equivalent) as in the square-pallet chain-pump, but the device would be set up vertically; and this indeed has long existed under the name of the 'paternoster pump' or 'rag-and-chain pump'. In this, the channel or flume must have a fourth side, and then, within this vertical pipe, an endless chain brings up balls of metal or lumps of rag (which nearly fill the pipe) to do the duty of pallets, bringing up water and discharging it at the top. The resemblance of the endless chain to a rosary gave the device its ecclesiastical name, and such pumps were popular for draining mines in Europe in the sixteenth century AD, though illustrations of it are known to have existed before this.

All this seems to have been foreign to traditional Chinese engineering practice, but this does not mean that the rag-and-chain pump may not have been a European invention deriving from the square-pallet chain-pump, possibly through vague hearsay from intermediate sources. The device had a period of importance in sixteenth-century Europe – at Schemnitz, in the Carpathians, such a paternoster pump was worked by ninety-six horses and raised water over 200 metres. This may have stimulated people to think about pistons in cylinders, but as the piston pump was familiar in Hellenistic and Arabic culture in the sixteenth century AD, it may have derived from the former rather than contributed to it. Provisionally, however, one might almost think of it as a marriage of the old European piston pump to a rumour from China. In any case the paternoster pump has today spread over the length and breadth of China.

The *sāqīya* (pot chain-pump)

The *sāqīya* ('the cup-bearer girl'?) or endless chain of pots differs in two ways from the chain of pallets in the inclined channel. First because it carries whole pots or buckets which fill at the lower end and discharge at the top, and secondly for the simple reason that the chain normally hangs straight down. A Babylonian relief of about 700 BC shows queues of men carrying baskets of earth upwards and descending with them empty, so it is not surprising that the chain of pots should have been an ancient idea. Philon of Byzantium (*c.* 210 BC) gives a description which seems at least partly genuine, while in 30 BC the Roman architect and engineer Vitruvius mentions the machine clearly. What is more, the remains of one of the chain-pumps used for emptying the underwater parts of the great ships of the Lake of Nemi built between AD 44 and 54 still exist. At all events, in the Hellenistic age (323 to 30 BC), it spread rapidly all over the Near East, and, in due course, the *sāqīya* or *daulāb* ('camel wheel') became as characteristic of the Islamic lands as the square-pallet chain-pump was of China.

That the machine was a comparatively late introduction to China is suggested by the name *gao zhuan tong che* [*kao chuan thung chhê*], the 'noria for high lifts'. It is first illustrated in AD 1313, but Fig. 505 shows the picture from The Exploitation of the Works of Nature of 1637. The significance of the name is that it suggests that the pots or bamboo pipes attached to the outer rim of the noria (p. 277), a machine with which the Chinese were already familiar, had flown away from it and taken their path on high to pass round a second wheel and so return. Illustrations we have of the high-lift *sāqīya* in China generally show a wheel at the lower end, but though this looks like a current operated paddle-wheel (Fig. 505), the texts all say that the drive came from above, using either a multiple pedal treadmill or a vertical wheel or whim driven by an ox. Moreover, the whole machine is shown slanting like a square-pallet chain-pump with a wooden guide-trough for the chain of bamboo buckets. Yet the Arab *sāqīya* was vertical, and did

高轉筒車

天工開物卷一

Fig. 505. A Chinese *sāqīya* from *The Exploitation of the Works of Nature* of AD 1637. Its foreign origin may be sensed by the inappropriate name, *gao zhuan tong che*, 'noria for high lifts'. It is unusual to have a wheel at the lower end, and one suspects operation by paddles, but the texts all make the power come from above.

not have a wheel at the lower end of the chain. The agriculturalist Wang
Zhen says that the 'noria for high lifts' was especially useful where water had
to be raised to considerable heights, and mentions lifts of several steps of
about 30 metres each. The fact that he names a particular temple, the Hu
Qiu Si at what is modern Wuxian, where such a machine was installed, may
mean it was rather uncommon. Indeed, the energy required may often have
made it uneconomic. Thus, it seems doubtful whether this device was ever
widely used in China; there are very few literary references to it, except in
the passages of agricultural encyclopaedias, few travellers describe it, and Dr
Needham never encountered it during his years in China during the Second
World War, though later, in 1958, he did see a few examples of some of its
parts discarded for more modern equipment. Nevertheless, one Chinese
industry, the salt fields at Ziliujing in Sichuan, adopted it at some date now
unknown.

The noria (peripheral pot wheel)

The noria (a word derived from the Arabic *al-nātūra*, 'the snorter') is the
most difficult of all these machines to trace back to its origin. It differs from
the square-pallet chain-pump and the *sāqīya* in that no chain is present and
the buckets, pots or bamboo tubes are attached to the circumference of a
single wheel, collecting water at the bottom and discharging it at the top.
Hence its Chinese name *tong che* [*thung chhê*] – 'tube machine'. This wheel
may be driven by the force of the current, if it is furnished with paddles, but
in still water it must of course be powered by men or animals. The first
Chinese illustration of it appears in the *Treatise on Agriculture* of AD 1313,
but Fig. 506 shows the semi-diagrammatic representation from the *Nong
Zheng Quan Shu* (Complete Treatise on Agriculture) printed in 1639. In
China, such norias, though built only of wood and bamboo, can attain
diameters of 13 metres or more, yet, even so, nothing larger than a small
stream is required to operate them. Sometimes, a number of norias are
arranged in a row (Fig. 507), but in the celebrated 15-metre diameter norias
on the Yellow River near Lanzhou in Gansu province, a battery of them is
arranged with their central shafts all in line (Fig. 508). These are of specially
stout construction to enable them to withstand the great river in spate.
Indeed, both in China and Indo-China, norias are often arranged in this
way, all of them on one single shaft.

It might be tempting to relate to the noria the strange phrase in the *Dao
De Jing* (Canon of the Dao and its Virtue), 'Some things are loading when
other things are tipping out', but in the fourth century BC this must surely
have referred rather to human chains of basket carriers. On the other hand,
the somewhat obscure account of Bi Lan's hydraulic constructions of AD
186, mentioned on p. 270, does seem to reveal the noria at work. Whatever

Fig. 506. The noria, or peripheral pot wheel; a drawing from the *Complete Treatise on Agriculture* of AD 1628. Like most traditional representations of the noria, this fails to do justice to the high lift available; the artist seems to have drawn the delivery flume in plan instead of elevation.

Fig. 507. A group of three norias at work near Chengdu, Sichuan. (Photo. Joseph Needham, 1943.)

Fig. 508. A battery of high-lift norias (diameter over 15 metres) in one of the arms of the Yellow River just below Lanzhou. Stoutly built, these can withstand, as here, the great river in spate. (Photo. Joseph Needham, 1958.)

it was that was built on the east side of the bridge seems to have been different from the square-pallet chain-pumps and siphons on the west. There is no doubt that the 'spread-eagled toad' means gear wheels of some kind, while commentators have taken *tian lu* [*thien lu*], which Dr Needham has translated as 'the heavenly pay-off' to mean some kind of animal, and hence part of the decorations on the machine. Though there is certainly this usage, he suggests that a pun may have been involved, for the top of the machine would certainly have been called the 'heavenly' end of it, and it was just here that the water poured out into the receiving channels. The use of gear wheels clearly indicates that the water was either still or, at any rate, not fast flowing, and indeed the Peace Gate at Luoyang was in the city's south-east wall near the bridge over the Yang Qu moat. Thus it seems most likely that Bi Lan's east-side equipment consisted of a battery of norias worked by men or animals, as well as perhaps square-pallet chain-pumps similar to those on the west side.

Distinct references to norias are rather rare in later literature, though one has the impression that they were quite widespread. The use of many large wheels for raising water for public baths, which could accommodate thousands of people, at Luoyang in AD 914, is referred to in an account of their builder, the monk Zhi Hui. A few decades later, an Arab traveller, Ibn Muhalhil, describes the water supply of the city of Shandan (see p. 301), which could hardly have been achieved without the use of one or more large norias, as in fact it was, virtually within living memory. Among other references, a pleasant literary one occurs in a travel book of about AD 1601, the *Yue Jian Bian* (A Description of Guangdong) by Wang Linheng. Speaking of the southern countryside, he says:

> As for the water-raising machines, at the end of every spoke there is a tube (bamboo bucket), upright when rising and inverted when descending, so that as the wheel turns, the water is poured out into a channel. The size of the wheel depends upon the height of the field – even as high as 9 or 12 metres a field can be watered. Not the slightest human effort is required. This is something like the water trip-hammers and water-mills of Zhejiang. Such machines as these for raising water would put to shame even the old man who lived north of the Han River long ago; in short the skill of man, when it comes to perfection, can conquer the works of Nature. Who invented (the noria) we do not know, but he ought to be honoured with worship and sacrifices.

As for the general history of the noria, it was known to Vitruvius (30 BC) and spread throughout the Hellenistic world, for it was certainly there in the second century AD, that is after Vitruvius and before Bi Lan. Though Vitruvius distinctly speaks of buckets, his noria is sometimes described as having a continuous series of radial boxes round the periphery, as if the very rim of the wheel were hollow and pierced with holes, a form which was long retained in the Near East. There was undoubtedly, therefore, a close relationship with the *tympanum*, a water-raising wheel completely boxed in and divided into compartments. However, this has the disadvantage that it only delivers water at axle level, and not at the top of the wheel, but it will do for small lifts. Nevertheless, the most splendid norias ever constructed were at Hama on the Orontes in Syria. Here, the largest had a diameter of over 21 metres, and discharged into tall arched stone aqueducts as well as the usual trestles, and all were originally constructed by Ibn 'Abd al-Ghanī al-Hanafī' (AD 1168 to 1251); the earliest Arabic literary reference is, however, of AD 884.

From the Islamic countries, the noria moved to Europe. It was particularly common in Spain; it was at work in France in the eleventh century AD; it appears in a German fifteenth-century manuscript; and it has persisted into modern times in some places – in Bulgaria, for instance, where the design is very Chinese. What is more, steam-driven norias, delivering over 9000 litres on each revolution, were at work in the 1890s for emptying copper residues into Lake Superior in the North American continent.

It seems, then, that we are presented by the noria with a problem similar to that which we shall encounter in the case of the water-wheel, namely first appearances separated by rather short time intervals at the two extremes of the Old World. The problems are, of course, closely related, since the noria operated by a current of water with paddles is very close in design to the water-wheel for mills. Since neither ancient Persia nor Greece seem to be the original home of the noria, the question therefore arises whether it might not be India. It was very widespread there in recent times, but the problem emerges of how far back it goes, due to the difficulty of dating Indian texts. However, there are references in Pali, a sacred language of the Theravāda Buddhist religious canon and of North Indian origin, to a turning wheel, which commentaries explain as a machine with water pots attached. If this is really the noria and not the *sāqīya*, the mention is interesting, since the date would be about 350 BC. Other early Indian languages, such as Classical Sanskrit as well as later ones, also provide references. Perhaps, too, comparisons in Buddhist literature between the noria and the Buddhist 'wheel of existence' point to an early use of the machine.

Summing up, the general impression gained is that the characteristic Chinese water-raising machine was the square-pallet chain-pump. Originating in the first century AD, it knew a worldwide diffusion after the sixteenth. The characteristic Hellenistic and Arabic water-raising machine was the *sāqīya* with its chain of pots; this was probably an early Alexandrian invention, and was transmitted to China from Arab lands some time before the fourteenth century AD. The noria is the most difficult to place, but provisionally we may say that it was invented in India, reaching the Hellenistic world in the first century BC and China in the second century AD.

THE USE OF THE FLOW AND DESCENT OF WATER AS A POWER SOURCE

So far, we have been concerned with people expending energy to move water about; now we approach the even more epic story of getting water to labour for them. The invention of the current operated noria had already done this, though only for the purpose of raising the water itself to a higher level. Who was it who first realised that the torque of the axle of the noria could be made to perform tasks beyond the power of unaided human strength, and with greater efficiency and continuity of effort than people or animals could bring? Or was the water supply wheel simply a kind of extension of the horizontally rotating millstone? This is not the place to elucidate all these problems of origin, so we shall just glance at the mass of facts relating to the invention and use of water-wheels in East Asia.

The term 'mill' must not be taken in the narrow sense only of the rotary quern (see this abridgement, volume 4, pp. 114 ff.). We shall have to deal with tilt-hammers and trip-hammers, edge-runner mills, sawmills, air conditioning fans, textile machinery and – surprisingly heading most of the others in date – metallurgical bellows operated by water-power.

Spoon tilt-hammers

Let us start with what amounts to a riddle. How could the power of falling water be made use of without any application of rotary motion? The answer is by a device which was the exact opposite of the swape (see p. 255). Instead of the bailing bucket being counterweighted to assist in raising water, the counterweight was turned into a hammer or pestle and alternately raised and allowed to fall by having a stream of water pouring into the bucket at the other end of the beam. This was the 'spoon tilt-hammer' or 'trough hammer', illustrated in all the books from the *Treatise on Agriculture* of AD 1313 onwards (Fig. 509). Unfortunately, references to it in literature are very few. The *Treatise on Agriculture* quotes a poem which is hard to identify, but there is a statement in a Ming *Zheng Zi Tong* (Dictionary of Characters) which says:

> The mountain people cut out wood in the shape of a spoon, and make it face a mountain torrent to work a water tilt-hammer. Sometimes it works slowly, but it quickly doubles the efficiency

槽碓

Fig. 509. Water-power in its simplest form; the tipping bucket as an intermittent counterpoise. The spoon tilt-hammer from the *Treatise on Agriculture* of AD 1313.

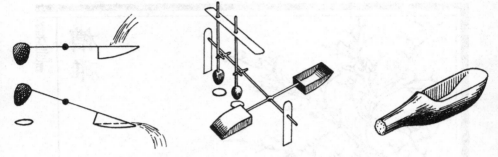

Fig. 510. Possible ancestry of the water-wheel. Left, the simple spoon tilt-hammer for pounding rice, etc.; right, its most archaic form, a hollowed-out log. In the centre, a double trough example working a vertical stamp-mill (in Settsu province, Japan). Because forms with four, six and eight spokes, with pallets and troughs, have been traditional, some connection with the water-wheel is not unlikely.

attainable by manpower. Ordinary people call it the 'spoon tilt-hammer'.

This would have been written in about AD 1600, though not printed until about thirty years later, but the device must have been well known already in AD 1145, for Lou Shou's poem (see p. 294) refers to the water flowing in and out of the slippery spoon. Before that we lose sight of it.

No references to the spoon tilt-hammer by travellers in China have been found, but late in the last century, J. Troup studied it carefully in Japan. Here, the *battari*, as it was called, had a trough with a sloping bottom to let out the water easily when the counterbalancing effect came into play, and in its most archaic form, as Lou Shou must have known it, it was simply a hollowed out log. In Settsu province, Troup saw a double trough which swung right round when emptying and rotated a shaft which worked vertical stamp-mill pestles by means of lugs (Fig. 510). Here the rotary principle had replaced that of the simple lever. He also found four-spoked examples in the Tonegawa Valley in Jōshū province. The two additional spokes were light and carried only flat boards with rims round the edges, and served only to help the main spokes get round so as to place the troughs under the stream of water. Troup also saw hammers where the spokes had been replaced by flat paddles, and even some with four, six or eight paddles ending in troughs. All the machines were operating stamp-mills. The relevance of these observations on the origin of the water-wheel we shall see shortly. But first we must sketch very briefly the essentials of what is known about the appearance and spread of the water-wheel in the West.

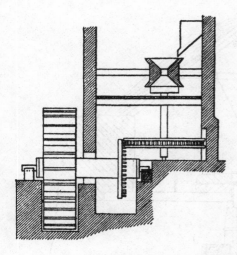

Fig. 511. The Vitruvian mill, a vertical water-wheel working the millstones by right-angle gearing.

Water-wheels in the West (and East?)

The oldest known water-mill (a water-wheel working rotary grinding stones for cereals) in the West was described by Strabo in about 24 BC as existing at Cabeira in ancient Pontus, then a Hellenised region of the Roman Empire lying just south of the Black Sea. The first literary evidence occurs in a Greek epigram dated about 30 BC. Then, in 27 BC, Vitruvius gave a matter-of-fact description illustrated in Fig. 511, and in about AD 74 Pliny gives an obscure reference which may be to stamp-mill hammers or to rotary querns. In the third century AD, there seem to have been mills along Hadrian's Wall and on the tributaries of the Moselle, and in the next century we begin to read of numerous mills, and the legal code of Justinian has much to say about them. The water-mill reached Germany in about AD 770 and England by 838 at least.

However, the mill of Vitruvius was unmistakenly a vertical 'undershot' wheel (where the water drives by passing under the wheel); the wheel itself being connected by right-angle gearing with the horizontally placed millstones. It is thus a vertical mill, but this is not the only possible arrangement, and over large parts of modern Europe, as well as the Europe we know from historical documents, it is clear that the only mills used were horizontal mills. In these, the water-wheel is mounted horizontally in the stream and connected directly to the millstones without gearing (Fig. 512). Possibly the mill in Pontus was of this type, but until the fifth century AD the only paintings and mosaics show undershot wheels. An overshot wheel (where the water pours down on the wheel from above) appears first in the fifth century AD.

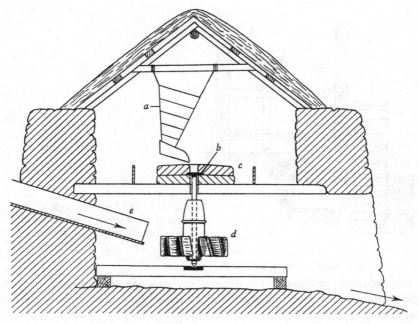

Fig. 512. The horizontal water-wheel, an example from the Isle of Lewis:
(*a*) hopper; (*b*) rynd; (*c*) stones; (*d*) wheel, or tirl, with obliquely set
paddles on thick shaft; (*e*) chute delivering the water to the side of the
wheel behind the shaft, as here viewed.

The early spread of the vertical water-wheel was mainly northwards, and
it became characteristic of France, Germany, England and Wales. But the
horizontal water-wheel seems to have moved along the lower margin of
countries, in the Lebanon and Syria, Israel, Thessalonica, Yugoslavia, Ru-
mania, Greece, Italy, Provence and Spain, then Ireland, the Shetlands, Scotland, the Faroes, Denmark, Norway and Sweden. What makes the pattern
still more interesting is that everywhere east of Syria – Iran, Turkestan, the
Unza region of the Himalayas – the horizontal wheel predominated. Moreover, it is the commoner form in Chinese illustrations. Whether or not the
Pontus mill was really horizontal, the invention was neither Greek nor Roman,
and the possibility presents itself that, together with the whole procession of
horizontal mills, its origins were further east. In any case, it is decidedly
simpler than the vertical mill-wheel because it needed no gearing, though
whether this is evidence of its primitive nature or of regression will be considered later. Here we need only add that in the horizontal wheels of most
regions, as modern observers have found them, the paddle boards of the

Fig. 513. Metallurgical blowing-engine worked by water-power; the oldest extant illustration (AD 1313) from the *Treatise on Agriculture*.

wheel are inclined (the angle is some 70°); otherwise, some form of scooping or cupping of the vanes is used to give a similar effect, as Troup found in Japan.

The metallurgical blowing-engines of the Han and Song

When we ask about the earliest water-wheel in Chinese history, we come upon the paradox that it was not used for turning simple millstones, but for the complicated job of operating blowing bellows for metallurgy (Fig. 513). This must mean that there was a tradition of millwrights going back some considerable time before, even though we cannot trace it in literary references. The essential texts run as follows; first the *History of the Later Han Dynasty* of AD 450:

> In the seventh year of the Jianwu reign period (AD 31) Du Shi was posted to be Prefect of Nanyang. He was a generous man and his policies were peaceful; he destroyed evil-doers and established the

dignity (of his office). Good at planning, he loved the common people and wished to save their labour. He invented a water-power reciprocator for the casting of (iron) agricultural implements.

[Commentary] Those who smelted and cast already had the push-bellows to blow up their charcoal fires, and now they were instructed to use the rushing of the water to operate it . . .

Thus the people got great benefit for little labour. They found the 'water(-powered) bellows' convenient and adopted it widely.

This advanced mechanism comes, therefore, between the dates of Vitruvius and Pliny. The tradition of Du Shi and his engineers must have persisted in Nanyang, for it was a Nanyang man who spread knowledge of the technique when he became prominent as an official two centuries later. This we know from the *History of the Three Kingdoms* of about AD 280, which says:

Han Ji when prefect of Luoling, was made Superintendent of Metallurgical Production. The old method was to use horse-power for the blowing-engines, and each *picul* [about 60 kilogrammes] of refined wrought iron took the work of a hundred horses. Man-power was also used, but that too was exceedingly expensive. So Han Ji adapted the furnace bellows to the use of ever flowing water, and an efficiency three times greater than before was attained. . . .

The period referred to must have been before AD 238. Some twenty years later, a new design seems to have been introduced by that ingenious man Du Yu. The story continues through the fifth century, and into the ninth.

Machines of this kind are illustrated in nearly all the relevant books from AD 1313, the time of Wang Zhen's *Treatise on Agriculture*, onwards. And not for metallurgical bellows only, but also for operating flour-sifters and any other machinery requiring a longitudinal or to-and-fro motion. Their historical importance is due to the fact that from some time in the first millennium AD they embodied the standard conversion of rotary to longitudinal reciprocating motion in heavy duty machines. For this reason, though it will disturb our pursuit of the cereal water-mill (p. 296), it seems necessary to deal with the problem now.

From the time of the earliest illustrations (see, for instance, Fig. 513), this form of conversion of rotary to longitudinal motion is present. At first sight, however, owing to many mistakes by the artist, it is almost impossible to make out from this how the machine worked. But by the aid of text, together with illustrations drawn in later books, a clear understanding of the mechanism can be reached; it is described in the caption to Fig. 514, where the use of an eccentric can clearly be seen. Now, we know that, during the

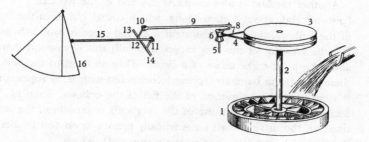

Fig. 514. Interpretation of Wang Zhen's horizontal hydraulic reciprocator for ironworks (blast furnaces, forges, etc.) and other metallurgical purposes (AD 1313). (1) Horizontal water-wheel; (2) shaft; (3) driving-wheel; (4) driving-belt; (5) subsidiary shaft; (6) smaller wheel or pulley; (7) eccentric lug or crank; (8) crank joint and pin; (9) connecting-rod; (10), (11) rocking roller bell-crank levers; (12) rocking roller; (13), (14) bearings; (15) piston-rod; (16) fan-bellows. We see here a conversion of rotary to longitudinal (to-and-fro) reciprocating motion in a heavy duty machine by the classical method later characteristic of the steam-engine, transmission of power taking place, however, in the reverse direction. Thus the great historical significance of this mechanism lies in its general paternity of steam-power. Both excursions of the bellows are fully mechanised in this design.

Song, the blowing apparatus in question was a large type of fan or piston bellows. Since a considerable increase in velocity was gained by having the eccentric mounted on a small wheel instead of the large one, and this was no doubt purposely intended, securing the main drive wheel by upper instead of lower bearings would have been deliberate. Now let us compare this with what Wang Zhen himself says in a most interesting passage:

> According to modern study (AD 1313!), leather bag bellows were
> used in olden times, but now they always use wooden fan
> (bellows). The design is as follows. A place beside a rushing
> torrent is selected, and a vertical shaft is set up in a framework
> with two horizontal wheels so that the lower one is rotated by the
> force of the water. The upper one is connected by a driving belt
> to a (smaller) wheel in front of it, which bears an eccentric lug
> (lit. oscillating rod). Then all as one, following the turning (of the
> driving-wheel), the connecting-rod attached to the eccentric lug
> pushes and pulls the rocking roller, the levers to left and right of
> which assure the transmission of the motion to the piston-rod.
> Thus this is pushed back and forth, operating the furnace bellows
> far more quickly than would be possible with man-power.

Another method is also used. At the end of the wooden (piston-) rod, about 1 metre long, which comes out from the front of the bellows, there is set upright a curved piece of wood shaped like the crescent of the new moon, and (all) this is suspended from above by a rope like those of a swing. Then in front of the bellows there are strong bamboo (springs) connected with it by ropes; this is what controls the motion of the fan of the bellows. Then in accordance with the turning of the (vertical) water-wheel, the lug fixed on the driving-shaft automatically presses upon and pushes the curved board (attached to the piston-rod), which correspondingly moves back (lit. inwards). When the lug has fully come down, the bamboo (springs) act on the bellows and restore it to its original position.

In like manner, using one main drive it is possible to actuate several bellows (by lugs on the shaft), on the same principle as the water trip-hammers. This is also very convenient and quick, so I think it worth recording here.

The first paragraph requires no comment, and can easily be understood from Fig. 514. The second system, however, though not illustrated in any of the books, is also quite significant, for it describes a type of blowing-engine which used only cams for the conversion of rotary to longitudinal motion. It depends on a vertical, not horizontal, water-wheel, and has been reconstructed in modern times (1959) by Yang Kuan. This is illustrated in Fig. 515, and is similar to what we shall see in the hydraulic trip-hammer (Fig. 517 below), where a series of hammers are alternately raised and allowed to fall by catches or lugs rotating with the main shaft. Since it was common in Han times, the blowers of Du Shi and Han Ji were surely of this kind, but there is always the difficulty that horizontally mounted mill-wheels seem in general to have been more characteristically Chinese than the vertical Vitruvian ones.

It is needless to emphasise the outstanding importance of power driven bellows and forges for the smelting of iron, and the remarkable early success of the Chinese in cast iron technology cannot be unconnected with the machinery just described. The hydraulic blowers of Du Shi and his successors had indeed a glorious future before them, for it is now clear that the use of water-power for metallurgical purposes began in Europe, notably in Germany, Denmark and France, very much later than in China. Forge-hammers were the first to be mechanised, at about the beginning of the twelfth century AD, and the use of water-power to drive the bellows used for the air blast followed early in the thirteenth century. Of the specific origin of the plans for European machines, we know nothing, but if anything came

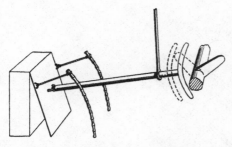

Fig. 515. Interpretation of Wang Zhen's hydraulic blowing-engine for ironworks (blast furnaces, forges, etc.) and other metallurgical purposes (AD 1313; after Yang Kuan). On the right, there is a shaft with a number of projections or lugs (only one is seen in the diagram) rotated by a vertical water-wheel, just as in the common water-powered trip-hammers (Fig. 517). During part of each revolution, the lug pushes back a crescent-shaped board at the end of the suspended piston-rod of a fan bellows. The return excursion is effected by bamboo springs, so that only one motion may be said to be mechanised from the main power source.

westwards overland it would have been the trip-hammer lug first, and then long afterwards the eccentric drive. The designs of the engineer Agostino Ramelli (1588) especially were strangely close to Song and Yuan Chinese patterns because of his extensive use of rocking rollers like bell-cranks. So also in the previous century we find an intimate resemblance in the sketches of Antonio Filarete (*c.* 1462) and the subsequent one in G. A. Böckler's *Theatrum Machinarum Novum* (1662). One cannot help believing that there was some underlying connection. In any case, the Chinese engineers seem to have had priority by some ten centuries for the practical trip-hammer principle, and three or four for the combination of eccentric, connecting-rod and piston-rod. Must we not also see in the latter the precise, but inverse, pattern of the arrangement of the reciprocating steam engine?

Reciprocating motion and the ancestry of the steam engine

All this brings up in rather acute form the history of converting rotary and reciprocating motions. The oldest examples of this achievement are devices such as the bow-drill and pump-drill (see this abridgement, volume 4, pp. 32 and 33), and these all involved non-continuous belting; in the development of machinery this did not lead very far. Next came the use of lugs on a rotating shaft, with springs to ensure return travel, though more commonly gravity had been used rather than springs, as in the vertical stamp-mills of Western medieval times and the ancient Chinese trip-hammers. Important also is the connecting-rod system of the Chinese blowing-engines, and a

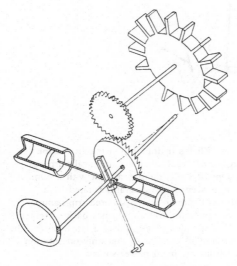

Fig. 516. Interpretation of the slot-rod force-pump for liquids by Aubrey Burstall. A gearwheel rotated by a vertical water-wheel engages with another beneath it. The lower toothed wheel is mounted on a shaft which is loosely pivoted at the right and free to rotate in an annular groove at the left, so that, while the wheel turns on its geometrical centre, the shaft describes a conical path and acts as an eccentric lug. Instead of being linked to any connecting-rod, however, it slides up and down within a slotted rod attached to a fixed pivot below and to two piston-rods, one on each side. Thus a continuous flow of liquid proceeds up the discharge pipe (not shown here).

force-pump, using a slotted rod devised by the Arab al-Jazarī in AD 1206, a century before the encyclopaedist Wang Zhen in his *Treatise on Agriculture* of 1313. But though al-Jazarī's machine (Fig. 516) was a fine device for the early thirteenth century, it was not as direct an ancestor as the Chinese system which Wang Zhen described.

Another machine which might be considered in close connection with the hydraulic blowing-engine is the silk-winding or reeling apparatus of the eleventh century AD (this abridgement, volume 4, pp. 69–70). This is because the main winding reels are worked by a crank-and-pedal motion, while the ramping arm for reeling the silk evenly is operated from the same power source by a driving-belt which rotates a pulley at the other end of the frame. This subsidiary wheel then moves the ramping arm back and forth by means of a lug placed eccentrically. Here then we have the driving-belt, just as in Wang Zhen's hydraulic blower, as also the smaller wheel with its eccentric, so that the ramping arm corresponds to the connecting-rod. There is nothing, however, corresponding to the piston-rod, which converts rotary to reciprocating motion. So the silk-reeler did not have quite all the components of the hydraulic blower. But the fact is that it was already a standard piece of equipment at the end of the eleventh century, and indeed may have been established practice long before owing to the antiquity of the silk industry. This strengthens the possibility that the water-powered blower, with its full 'steam engine' arrangement for converting rotary motion to a longitudinal to-and-fro motion, developed during the Tang and Song, that is four or five centuries before Wang Zhen's description of it. It may also therefore be older than al-Jazarī's swaying slot-rod.

It must be emphasised that the system of three parts – eccentric, connecting-rod and piston-rod – has not so far been found in any fourteenth-century European illustration, and occurs only rarely in the fifteenth century. Indeed, when in the late fifteenth century, nearly 200 years after Wang Zhen, Leonardo da Vinci faced the problem of converting rotary to reciprocating motion, he displayed a most curious disinclination to use the combination of eccentric (or crank), connecting-rod and piston-rod. He did so only for a mechanical saw.

Perhaps the most extraordinary part of the whole story is that James Watt was driven to the invention of his 'sun-and-planet' gear because the basic method of converting rotary and reciprocating motion by eccentric and piston-rod had been patented for steam engines by James Pickard in 1780. Watt had not patented it himself because he knew that it was old, but probably no one involved knew of the fifteenth-century German engineers, and certainly no one then could have had any suspicion that the Chinese of the Song period had been intimately and practically acquainted with it. Indeed, on our present information, they were its real inventors.

It only remains to add a very brief word about the coming of the steam-engine to China in the nineteenth century. The steamboat was the carrier. It has generally been thought that the East India Company's steamer *Forbes* was the first to arrive, in 1830, with the invention. But, interestingly, a passage in the *Hai Guo Tu Zhi* (Illustrated Record of the Maritime Nations) of 1844 shows that it was in April 1828 that it entered the scene in China, for '. . . there suddenly came from Bengal a "fire-wheel boat"', which it then describes.

Hydraulic trip-hammers in the Han and Jin

Of all the different types of machines driven by water-power in early times in China, that which is most mentioned in the literature is the trip-hammer (*shui dui* [*shui tui*]) (水碓). This was the simple mechanism of the pedal tilt-hammer in which the hammers were operated by a series of catches or lugs on the main revolving shaft. All books from AD 1300 illustrate this, but reproduced here is the drawing from the *Explanations of the Works of Nature* of 1637 (Fig. 517).

An important point to note is that, while the horizontal water-wheel is much the simplest arrangement for rotary millstones, the trip-hammer is best suited by the vertical water-wheel, and this is the type usually depicted. One cannot help wondering, however, whether many of the early machines now to be mentioned were not horizontal water-wheels with right-angle gearing. Another point is that, in Chinese practice, the hammers were always recumbent, and not vertically acting stamp-mill pestles such as we find in medieval Europe; this made possible a considerably heavier installation in China.

The earliest explicit statement seems to be in the *Xin Lun* (New Discourses) of Huan Tan (*c.* AD 20), who remarked:

> Fu Xi invented the pestle and mortar, which is so useful, and later
> on it was cleverly improved in such a way that the whole weight of
> the body could be used for treading on the tilt-hammer, thus
> increasing the efficiency ten times. Afterwards the power of animals
> – donkeys, mules, oxen and horses – was applied by means of
> machinery, and water-power too used for pounding, so that the
> benefit was increased a hundredfold.

Words so general and so assured authorise the conclusion that from at least the time of the emperor Wang Mang (*reigned* AD 9 to 23) onwards water-wheels were used more and more for working pounding machinery. Other Han references confirm this, and in the third and fourth centuries AD references are abundant. In about AD 270, Du Yu, a high official and engineer, established 'combined' trip-hammer batteries, which probably means that several shafts were arranged to operate off one large water-wheel. Indeed, there were many men famous for possessing many such machines, even hundreds of them, such as Shi Chong (*died* AD 300), who operated them in more than thirty districts. At one time, they were so numerous that there were even regulations preventing them being set up within a certain radius (50 kilometres) of the capital.

Poems were written about them, and among the many references during the Tang and Song was this one in about AD 1145, in which the poet and official Lou Shou wrote:

> The graceful moon rides over the wall
> The leaves make a noise, sho-sho, in the breeze.
> All over the country villages at this time of year
> The sound of pounding echoes like mutual question and answer.
> You may enjoy at your will the jade fragrance of cooking rice,

水碓

盍利
芽用

Fig. 517. A battery of hydraulic trip-hammers worked by an undershot vertical water-wheel, from *The Exploitation of the Works of Nature*, AD 1637.

Or watch the water flowing in and out of the slippery spoon.
Or listen to the water-turned wheel industriously treading.

During the course of time, moreover, the water trip-hammer was put to many uses other than hulling rice. As we have seen, it was used in forges, and it is interesting to find a Daoist connection, for the pharmaceutical hermits employed it for crushing mica and other minerals as drugs. Then, in the Ming, we have an account of the use of trip-hammers by the paper-makers of Fujian, while in the Qing, about AD 1600, it was said of the perfume-makers of Guangdong that the fragrance from their mills was often carried away downstream for kilometres.

Once more we may note that the trip-hammer was the direct ancestor of all heavy mechanical hammers until the time of the introduction of the steam-hammer in AD 1842. Indeed, in the eighteenth century, western types of forge-hammer reproduce the Chinese design systematically with hardly any modification. Yet the trip-hammer was not generally seen in Europe until it appears in a woodcut of a forge-hammer in 1565 in a book by Olaus Magnus, though Leonardo da Vinci also sketched it before this. The forge-hammer goes back earlier than this, at least to 1190, but it was probably of the form of the lighter vertical stamp-mill of medieval Europe. Perhaps there may be some significance in the fact that the other great use of the tilt-hammer operated by water-power was for compressing or fulling cloth, since there is reason for thinking that much Chinese textile machinery design made its way to Europe in about the time of Marco Polo (late thirteenth century AD).

Water-mills from the Han onwards

It is curious that references in early Chinese literature to the mill *par excellence*, the rotary millstones driven by water-power, are much rarer than those to the water-driven trip-hammer. This may arise perhaps from a fluidity of technical terms at that time, especially among scholars whose qualifications were not technical. It would not have been difficult to confuse *chui* [*chhui*] (硾), the equivalent of the grain-mill proper, or even *wei* (磑) with the word *zhui* [*chui*] (碓) (hammer) used for *dui* [*tui*] (碓), the tilt- or trip-hammer. Moreover, in the texts already referred to, some books and some editions write *wei* or *mo* (磨) instead of *tui*. Thus, multiple mills worked from a single water-wheel by gearing were attributed to Du Yu (AD 222 to 284); and water-mills instead of water-powered trip-hammers to Chu Tao (c. AD 240 to 280) and Wang Rong (AD 235 to 306). There seems, in effect, little reason to doubt that water-driven quern mills were working at least as early as the hydraulic blowing-engines of the first century AD, and perhaps some time before then.

The first water-mill illustrations in the main picture tradition start from AD 1313, and the picture of a horizontal water-wheel from the *Treatise on Agriculture* of that date is shown in Fig. 518. Another (Fig. 519) shows an overshot vertical water-wheel driving from six to nine mills geared together by toothed wheels. Wang Zhen tells us that in his time there were some installations so large that trip-hammers and edge-runner mills as well as millstones were all worked by shafts and gearing from one great water-wheel. Where conditions suggested it, the main water-wheel was also equipped as a noria for raising water in times of drought. Some of these combined factories, as we must call them, could mill enough grain daily for 1000 families. Wang Zhen found, when he travelled in Jiangxi, that plant of this kind was widely used for the pounding and rolling of tea leaves, and from other sources we know that there were one hundred such tea mills in AD 108, and more than 260 in 1079.

A whole technical vocabulary must have been used by these Chinese millwrights, as by those in the West. But in China some such water-wheels still exist, as two of Dr Needham's photographs from 1958 illustrate (Figs. 520 and 521). He also remarks that while in Gansu province he easily found mills of Vitruvian style with right-angled gearing.

Celebrated water-mills, which the emperor himself came to inspect, were erected by the great mathematician and engineer Zu Chongzhi in about AD 488, and in about 600 Yang Su, one of the chief technologists of the Sui dynasty, was in control of, or owned, thousands of them. As time went on, however, conflict between the interests of irrigation controllers and those who used water-power greatly increased until, from the Sui onwards, officialdom came into head-on collision with a developing use of shipping. They therefore decided that the water-mills must not interfere with water-conservancy, and this was explicitly laid down in the Tang dynasty ordinances dating from AD 737. Thenceforward, certain officials distinguished themselves as persecutors of millers. Indeed, the largest destruction of millwrights' work occurred in 778, when eighty plants were torn down, not excluding two water-wheels owned by the general Guo Ziyi, who had saved the empire from a serious rebellion. However, as the mills were generally the property of imperial concubines, powerful eunuchs or rich merchants, and also one of the richest sources of income for the great Buddhist abbeys, so the opposition of the Confucian bureaucrats was really only one aspect of a perennial antagonism.

During the Tang, the water-mill had radiated to other countries in the Chinese culture-area, to Japan (via Korea) in AD 610 and 670, and to Tibet in about 641. Later on, peoples such as the Qidan Tartars were quite familiar with it. Then, early in the tenth century AD, the abundance of water-mills in China caught the attention of an Arab traveller, Abū Dulaf Misʿar

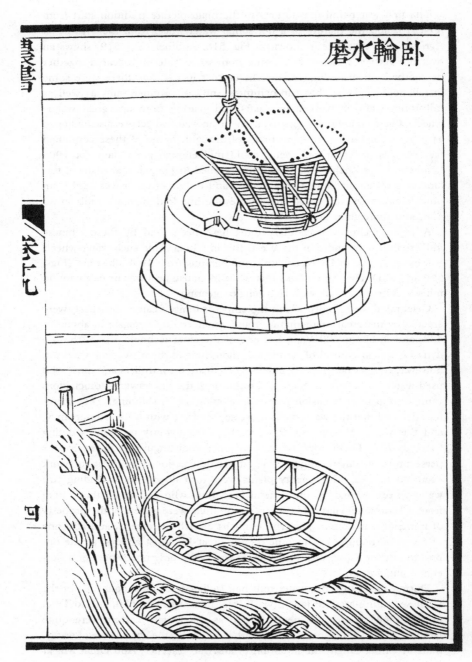

Fig. 518. The most characteristic Chinese form of water-mill, a millstone driven by a horizontal water-wheel; from the *Treatise on Agriculture* of AD 1313.

Fig. 519. Geared water-powered milling plant, nine mills being worked by an overshot vertical water-wheel and right-angle gearing, an illustration of AD 1742 in the *Complete Investigation of the Works and Days*. The oldest picture of such a plant is in the *Treatise on Agriculture* of AD 1313.

Fig. 520. Horizontal rotor, still unfinished, at Chengdu. (Photo. Joseph Needham, 1958.)

Fig. 521. Horizontal water-wheel in position under a water-mill near Tianshui, Gansu, but not working at the time. (Photo. Joseph Needham, 1958.)

ibn al-Mulhalhil, who described no less than sixty mills on canals in and around what he believed to be the capital city of Sandabil, which seems likely to have been Shandan in Gansu province on the Old Silk Road. This had some remarkable waterworks, which probably existed until late in the Qing dynasty. Perhaps Muslim ritual hygiene had exerted a stimulating influence on this frontier region earlier than ibn al-Mulhalhil's time, possibly in some other Chinese cities as well.

There remains little more to say. From the Tang references mentioned, it is clear that the edge-runner mill so prevalent in China (see this abridgement, volume 4, pp. 124–6) should have been powered by water from an early time. Indeed, their origin seems fixed rather definitely between AD 390 and 410, for in the biography of Cui Liang who lived at this time, we read:

> When (Cui) Liang was in Yongzhou, he read the biography of Du Yu, from which he learnt that Du had devised the 'eight (geared) mills', greatly benefiting his contemporaries thereby. Cui therefore taught the people (to apply water-power to) edge-runner mills and roller mills. After he had attained the position of Grand Counsellor, he memorialised the emperor suggesting that a dam should be built on the Gu Shui east of the Zhangfang bridge (to provide water) for water-powered runner mills and roller mills. So these were established in several tens of places, and the profit to the country was ten times greater than ever before.

In AD 550, the first emperor of the Northern Qi presented a 'set' of edge-runner mills to the dethroned emperor of Eastern Wei. Although this type of mill shared with others the vicissitudes of government regulation in the Tang, it persisted virtually unchanged until modern times. Indeed, Dr Needham found many at work still in Sichuan province; generally with only one vertical grinding stone. It also found its way to Europe, where it became known as the 'gunpowder mill', and held its own until the coming of steam and electric power.

Lastly, we must glance at a few additional uses of water-power. Sawmills powered in this way have been considered an early development in Europe, one such being allegedly referred to in AD 369, and a famous one illustrated about 1237. Specifications for such machines appear in the *Qi Qi Tu Shuo* (Diagrams and Explanations of Wonderful Machines) of AD 1627, but it would not be safe to assume that the idea had never occurred to anyone in China previously, though we have no evidence of it. Water-wheels were there put to all sorts of uses, such as polishing stone columns for buildings. Another remarkable use was in AD 747 during the Tang for working air-conditioning fans.

Particularly remarkable was the use, at least as early as AD 1313, of water-power for textile machinery. The *Treatise on Agriculture* illustrates an undershot water-wheel and a large driving wheel with a belt drive on the same shaft working a multi-bobbin spinning machine for hemp and ramie, perhaps also for cotton. This should be enough to give pause to any economic historian, especially as Wang Zhen clearly says that such installations were common in his time in districts which grew these textile crops. Indeed, traditional Chinese culture developed a multifarious use of water-power, so that in AD 1780 the Korean visitor Pak Chiwǒn, remarking on water-power being used for blowing air for furnaces and forges, winding off silk from cocoons and milling cereals, concluded that 'there was nothing for which the rushing force of water to turn wheels was not employed'.

The problem of inventions and their spread

When we survey the whole of the above evidence, we may wonder whether perhaps the horizontal water-wheel and the vertical water-wheels were not two entirely distinct inventions. On this provisional conception, the vertical water-wheel would have been an adaption of the (originally Indian?) noria, while the horizontal wheel would have been, as it were, a downward extension of the running component of the rotary quern. The right-angle gearing required by the Vitruvian type might be considered primarily Alexandrian, for though the Han engineers were quite expert with gear wheels, the Alexandrians were perhaps just a little ahead of them. On this view, the originally Chinese horizontal water-wheel would have made its appearance in Persian Pontus under Mithridates VI (123 to 63 BC), and then continued its spread, undeterred by the Vitruvian design, around the coasts of Europe to end up in Scandinavia as the 'Norse mill'. The westward and northward spread must have taken place in the first Christian centuries.

As for the comparative dating of water-power between China and the West, it clearly constitutes a case of approximately simultaneous adoption as puzzling as that of rotary milling itself. The Asia Minor date of 70 BC or so is authorised by subsequent writings of the same century, but the first century BC for China is inescapably indicated by the words of Huan Tan written in AD 20, though in which of its decades a water-wheel was first made to drive the trip-hammers there remains obscure. The first century BC is also indicated by the fact that the earliest specific description in a Chinese source refers to metallurgical furnace blowing, a job more complicated than one would think than the simple grinding of cereals, and hence implying a longer period of development.

The vertical water-wheel might seem at first sight to have had the greater future, but in fact it was the other way round. For the horizontal water-wheel was the direct ancestor of one of the most impressive power-sources of Post-Renaissance times, the hydraulic and steam turbine. This is because, in the horizontal wheel, the movement results wholly from the impact or impulse of the water acting on the vanes, just as in the turbine (where steam replaces the water). Vertical overshot wheels, however, are turned mainly by the weight of the water rather than its momentum. There was, of course, Heron of Alexandria's 'aeolipile', which rotated due to a reaction of steam issuing from jets. Thus the turbine is essentially a combination of the ancient Chinese water-wheel with the Alexandrian aeolipile.

By now the reader may have had enough of mills, but there remain two kinds about which nothing has yet been said. One was the water-wheel drive for a clock (already discussed in this abridgement, volume 4, chapter 6); the other was the mill mounted in boats. In other words, all paddle-wheel ships are children of the water-mill, and China was the land of their infancy. To these we now turn.

WATER-WHEELS DRIVEN AND DRIVING; SHIP-MILLS AND
PADDLE-BOATS IN EAST AND WEST

The first story begins, unlike most stories in this book, in Rome. In the year AD 536, the Goths were besieging the city, and they cut off the water of the aqueducts which drove the Janiculum mills on the right bank of the River Tiber. It was not possible to work them with animals for the city was short of food, and provision could scarcely be found for the horses, so the defenders were on the point of being reduced to starvation. However, the Byzantine general, Belisarius, who commanded the garrison, conceived the idea of mounting the mills with their water-wheels on boats moored in the Tiber.

It would seem that this system was afterwards very widely used. Ship-mills were in Venice in the eleventh century AD, and in France from the twelfth to the end of the eighteenth, but made a brief appearance in England only during the sixteenth. By the late fifteenth century, these floating mills were depicted in an illustrated manuscript as being in Rome itself, using very broad water-wheels, while on the Dniester and Danube rivers they continued down till contemporary times.

It has been recognised that ship-mills were not confined to Europe, and though there seem to be no Chinese descriptions as early as the one from Belisarius, their employment can hardly have been much later. This results from the fact that the Ordinances of the Tang Department of Waterways, dating from AD 737, forbid ship-mills on the river and streams near Luoyang as if they were something very well known. Other early references are still scarce, but mills on boats were mentioned by the poet Lu Yu in a poem written in Sichuan in 1170. Then, in about 1570, Wang Shimou describes in his *Min Bu Si* (Records of Fujian) how the paper-makers there mounted their trip-hammers on boats each with two water-wheels and furiously pounded away by the aid of the fast flowing current of their rivers. So also

Wang Shizhen, writing about his own journey to Sichuan in the early Qing period (early seventeenth century), saying that in Liangjiang there were many ship-mills carrying out grinding, pounding and sifting by the use of water-power. He even commented that the boats made a noise 'ya-ya, ya-ya' incessantly.

Another visitor was Robert Fortune, who in 1848 travelled through the tea country of northern Fujian. While still in Zhejiang province, he found a whole colony of ship-mills near Yanzhou:

> Leaving the town of Yanzhou behind us, our course was now in a north-westerly direction. The stream was very rapid in many parts, so much so that it is used for turning the water-wheels which grind and husk rice and other kinds of grain. The first of these machines which I observed was a few miles above Yanzhou. At the first glance I thought it was a steamboat, and was greatly surprised; I really thought that the Chinese had been telling the truth when they used to inform our countrymen in the south that steamboats were common in the interior. As I got nearer I found that the 'steamboat' was a machine of the following description. A large barge or boat was firmly moored by stem and stern near the side of the river, in a part where the stream ran most rapidly. Two wheels, not unlike the paddles of a steamer, were placed at the sides of the boat, and connected with an axle which passed through it. On this axle were fixed a number of short cogs, each of which, as it came round, pressed up a heavy mallet to a certain height, and then allowed it to fall down upon the grain placed in a basin below. These mallets were continually rising and falling, as the axle was driven rapidly round by the outside wheels, which were turned by the stream. The boat was thatched over to afford protection from the rain. As we got further up the river, we found that machines of this description were very common.

Thus in Zhejiang province, if Fortune's account is right, water trip-hammers were the chief instrument used. The statements of Lu Yu and Wang Shizhen, however, both refer to the Yangtze in eastern Sichuan, and their descriptions, though written so long ago, can be complemented by that of a twentieth-century maritime historian, G. R. G. Worcester, who has given excellent engineering drawings of the ship-mills still in action around the city of Fuzhou, some ninety-six down-river from Zhongjing (Fig. 522). In this place, they carry four water-wheels on two axles. Those seen by Worcester carried treadle-sifters as well, but evidently in former times, or elsewhere, these were connected to the source of power, as shown in books such as the *Treatise on Agriculture*.

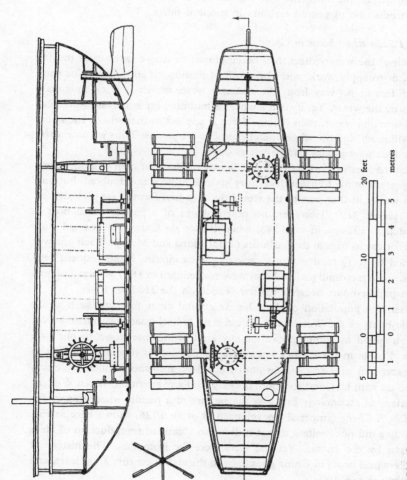

Fig. 522. Scale drawings of one of the ship-mills at Fouling, Sichuan. Each mill is worked by two stoutly built paddle-wheels and right-angle gearing. Two treadle-sifters are mounted within the hull. (After Worcester (1940).)

20 feet

metres

0   1   2   3   4   5

0   10   20

5   0

A minor point of considerable interest recorded by Worcester is that the power-shaft gears had eighteen teeth, while the quern gear-wheel had only sixteen. The system may therefore have been a step towards the modern engineering practice of introducing a 'hunting tooth' to ensure even wear. This principle did not come into regular use in marine engineering before the introduction of geared turbines in modern times.

Paddle-wheel boats in China

Of course, the water-wheel may be employed to derive work from moving water, or to apply work, with the result of motion, to still water. The ship-mill differs in no way from the ordinary water-wheel; it is ex-aqueous – driven by the water. Yet if such a wheel is mounted on astructure which can travel over the water, then if some force is applied to that wheel, the structure will move and the wheel then becomes ad-aqueous. Thus we now arrive at the true ad-aqueous paddle-wheel boat.

That the idea of working such paddles by the force of men or animals was fully present in the fourteenth century AD in Europe is not doubted, but how much earlier in that part of the world the idea was formed we shall discuss later (see p. 319). However, no practical use of a paddle-wheel boat is recorded in Europe in AD 1543, when Blasco de Garay constructed such boats for use as tugs in the harbours of Barcelona and Malaga. Each one was manned by forty men working capstans or treadmills. In subsequent centuries, some treadmill paddle-boats were made, until in 1807 Robert Fulton's steam paddle-boats began a regular service on the Hudson River.

When the population of the Chinese coastal cities first saw such steam paddle-boats of the Westerners, an old term 'wheel-boat' was remembered, and remained in common use for any kind of steamer down to our own times. People in these cities knew little or nothing of their own past, with the exception of a few old-fashioned scholars, for whom much fable was mixed up with fact, and to whom nobody paid any attention. When Western historians of technology first saw the picture of a paddle-wheel ship in the *Tu Shu Ji Cheng* (Imperial Encyclopaedia) of AD 1726, shown here as Fig. 523, they did not hesitate to put it down to a garbled reproduction of ideas brought by the Jesuits. Yet the facts were far from this. The history of paddle-wheel boats in China goes back to the eighth century AD at least, and probably to the fifth.

Naturally the earliest references are less clear than the later. But the original inventor may well have been the famous engineer and mathematician Zu Chongzhi, for in both his biographies there is mention of a 'thousand-league boat', which was tested on the Xinting River, south of modern Nanjing and proved to be capable of making several hundred *li* in one day without help of wind. This invention was made between AD 494 and 497, not long

車輪舸圖

Fig. 523. The paddle-wheel warship in the *Imperial Encyclopaedia* of AD 1726, but copied from the *Treatise on Armament Technology* of AD 1628. The four paddle-wheels are worked by treadmills within the hull.

before his death in 501. It is possible, however, that protected treadmill paddle-wheel boats may already have been used at the beginning of the fifth century, in a naval action under the command of Wang Zhene, who, as we have seen, was one of the admirals of the Liu Song dynasty. In his biographies, we find the following passage:

> Wang Zhene's forces sailed in covered swooping assault craft and small war-junks. The men propelling the boats were all (hidden) inside the vessels. The Qing (barbarians) saw the ships advancing up the Wei (river) but could not see anyone on board making them move. As the northerners had never encountered such boats before, every one of them was sore afraid, and thought that it was the work of spirits.

The passage goes on to describe how, after being moored, the boats cast off upon orders being given and moved away, apparently by themselves. Since the barbarians would surely have been familiar with oars and sails, something else is rather strongly suggested. In any case, the date is certain; the action took place in AD 418.

The following century gives many further indications of paddle-wheel boats. In AD 552, an admiral of the Liang dynasty, Xu Shipu, in the course of his campaign against Hou Jing, constructed a number of different kinds of craft to strengthen his fleet. The lists mention various types, including *shui che* [*shui chhê*] (water-wheel boats). While the usual meaning of this last term is, of course, to water-raising square-pallet chain-pumps, the context here (a list of boats) shows that a vessel of some kind was intended, and the obvious possibility is that paddle-boats were in fact meant. This same expression for a boat occurs again in the *Jing Chu Sui Shi Ji* (Annual Folk Customs in the Regions of Jing and Chu) by Zun Lin, where it is said that on the fifth day of May the river-people hold races with these 'water-wheel boats'. This book is often ascribed to the Liang period, about AD 550, and is certainly not later than the beginning of the Tang (*c.* 620). The same boats, it says, were also called *shui ma* (water-horses). An alternative interpretation, equally tenable, would take *shui che* as 'water-chariot', and suppose that all the boat names in this entry of Zun Lin's book on the folk festivals referred simply to the usual many-crewed hand-paddled dragon-boats in which people raced each summer. The classical confusion between *che* as 'chariot' and *che* as 'machine' renders the problem insoluble pending further evidence. However, another commander in the campaign against Hou Jing, the admiral Wang Sengbian, is said to have had in his fleet 'ships which had two dragons on the sides to enable them to go very fast'. The words 'two dragons' (*shuang long*) may have been a literary emendation for *shuang lun* 'two wheels', a proposal all the more plausible in that the context says a good deal about portents which appeared to the armies. Some twenty years later, at the siege of Liyang in AD 573, when the Northern Qi State was being invaded by the Chen, a fourth admiral, Huang Faqiu, who was also a distinguished military engineer, built and used a number of 'foot-boats' with success, and these can hardly have been anything else than treadmill-operated paddle-wheel boats.

The circumstantial evidence, therefore, seems distinctly strong that the original invention was made about the time of Zu Chongzhi towards the end of the fifth century AD if not a little earlier. When we come to the time of Li Gao, prince of the Tang, there can be no further doubt. His experiments with paddle-wheel boats were made between AD 782 and 785, when he was Governor of Hongzhou.

> Li Gao, always eager about ingenious machines, caused naval vessels to be constructed, each of which had two wheels attached to the side of the boat, and made to revolve by treadmills. These ships moved like the wind, raising waves as if sails were set. As for the method of construction it was simple and robust so that the boats did not wear out.

So far the *Old History of the Tang Dynasty* of AD 945; the *Xin Tang Shu* (New History of the Tang Dynasty) of 1061 adds that the prince himself taught his artisans how to make these craft, and that their speed was faster than that of a charging horse.

After this, it would be natural to find various echoes of practical paddle-boats in the literature, and so there are. However, it was not until the beginning of the Southern Song, early in the twelfth century AD, that the tread-mill paddle-wheel ships really came into their own. After the loss of the capital Kaifeng in 1126, the move of the Song administration to the southern provinces led to the first establishment of a regular Chinese navy based on the maritime expertise of the south. The Yangtze, it was said, must now be China's Great Wall, and battleships must be her watch-towers. Response to this stimulus was quickly apparent, for certain 'flying eight-bladed paddle-wheelers' helped the Song general Han Shizhong to inflict great losses on the Jin Tartars in 1130 when they were retreating north across the Yangtze. This set the pattern for the next century. Indeed, two years after the victory, the engineer Wang Yanhui, memorialised the emperor, saying

> To defend the thousand-*li* vastness south of the Great River, it is necessary to have warships; to halt the horsemen of the northern plains one must have vehicles. . . . Boats and vehicles are best if they are light and fast. . . . I have designed a 'flying tiger warship' with four wheels at the sides. Each wheel which has eight blades, is rotated by four men. This ship can travel a thousand *li* a day.

This description of 1267 is of particular interest because it is the oldest we have which mentions four wheels, exactly as depicted in the *Wu Bei Zhi* (Treatise on Armament Technology) of AD 1628 and the *Imperial Encyclopaedia* of 1726.

As it happened, however, the real proving ground for the developing invention was not the war against the northerners but strife within the Southern Song itself. During the Jin invasions there was a peasant revolt. Well captained by Yang Yao and Yang Qin, the rebels made themselves masters of the Dongting Lake and incessantly raided the cities on its shores. In the following year, the governor of Dingzhou embarked on a big shipbuilding programme to defeat them, including many junks with paddlewheels designed by a remarkable engineer. From a contemporary Dingzhou writer, we learn that

> ... one of the soldiers, Gao Xuan, who had formerly been Chief Carpenter of the Yellow River Naval Guard Force, and of the Baipo Vehicular Transport Bureau of the Directorate of Waterways, submitted a specification for wheeled ships which (he claimed) could cope with the enemy. . . . (He first built) an eight-wheel boat as a model, completing it in a few days. Men were ordered to pedal the wheels of this boat up and down the river; it proved speedy and easy to handle whether going forward or backward. It had planks on both sides to protect the wheels so that they themselves were not visible. Seeing the boat move by itself like a dragon, onlookers thought it miraculous.
>
> Gradually the number and size of the wheels were increased until large ships were built which had twenty to twenty-three wheels and which could carry two or three hundred men. The pirate boats, being small, could not withstand them.

This was a truly remarkable piece of technology, and an attempt to capture its flavour is shown in a reconstruction (Fig. 524). It is almost surprising that these craft were not called 'centipede-ships', certainly no other civilisation produced anything like them. But the plan was a very rational one, for in the absence of steam-power and cast-iron wheels it was necessary to distribute the strain over a larger number of paddles. Another source tells us that the biggest wheel-ships of the Dingzhou Governor were 60 to 90 metres long and capable of carrying between 700 and 800 men.

Very soon, these ships were ploughing the waves in the revolutionary service, for a government fleet of twenty-eight sea-going junks and two eight-wheel paddle-boats was stranded in a tidal river, so that all were captured and Gao Xuan himself taken prisoner. The *Ding Li Yi Min* (Recollections of Dingzhou) of about AD 1150 continues:

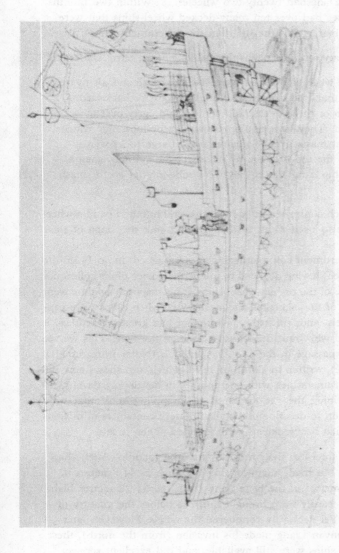

Fig. 524. Reconstruction of one of the multiple paddle-wheel warships of the Song period (*c.* AD 1135) based on the designs of Gao Xuan. The largest of these vessels had, as here, twenty-two paddle-wheels (eleven on each side) and a stern wheel. Joseph Needham writes: 'We have assumed the characteristic junk build, and have added auxiliary sails, a deck-castle, and a number of manned trebuchets (catapults) for hurling gunpowder bombs, poisonous-smoke containers, etc. The general's pennant flies astern, and a banner hoisted amidships says "Support the Song, destroy the Jin!"' Such men-of-war carried crews of 200 or 300 sailors and marines. In the drawing, the housing has been removed to show the forward six paddle-wheels on the port (left) side.

The pirates thus secured the design of the paddle-wheel boat and also the chief designer. He built for Yang Yao a large ship of the Hezhou style with several decks and twenty-four wheels and, for Yang Qin, a Dadeshan twenty-two wheeler. . . . Within two months the pirate bases had over ten many-decked wheel-ships that were stronger and better constructed (than the government ships).

Another contemporary source explains the meaning of the term 'Dadeshan'.

In the paddle-wheel ships men were stationed fore and aft to tread on pedals so that (the vessels) could go forward or backward. . . . They had two or three decks, and some could carry over a thousand men. They were equipped with 'grappling-irons' which were like great masts over 30 metres high. Large rocks were hoisted up to the top of these by means of pulleys and when a government ship came close, they were suddenly let go to smash her.

The rebel fleet, which comprised at its height several hundred paddle-wheel ships of all sizes, also used rams to damage and sink the ships of their opponents.

Conservative government commanders were nonplussed; in AD 1135, the rebel chief was routed, having declined to follow the advice of his colleagues and to make full use of the paddle-wheel ships at his disposal. As time went on, however, some of the eighteen- and twenty-two-wheel ships were captured from the rebels, and, since the government had greater resources, it was they in the end who built the largest vessels of the type. This we see from an interesting passage in the *Lao Xue An Bi Ji* (Notes from the Hall of Learned Old Age), written in about AD 1190, which also shows how the rebels and the government vied with one another in developing naval techniques. After describing the various kinds of 'grappling-irons' that were used, the text goes on to describe thin walled pottery containers of noxious chemicals, with which heavy defeats were inflicted. Then it says

The imperial forces in their turn imitated the (paddle-wheel) ships of the rebels, but made them larger – as much as 110 metres in length, 12.5 metres in the beam, and (with masts) 22 metres high. But they had hardly been brought into use before the infantry of (general) Yo Fei decisively conquered the rebels. However, later on, when Wanyan Liang made his invasion (from the north), these paddle-wheel ships were still available, and did excellent service.

These last two remarks need a little explanation. The rebels were finally defeated by the famous commander Yo Fei, who caught the greater part of

their fleet in a strange ambush. Having covered the water in an arm of the
lake with masses of floating weeds and rotten logs, he lured them in, and
when the paddle-wheels were all entangled so that they could not move, his
boarding-parties swarmed on to the ships and won a decisive victory. Yang
Yao himself was killed. This was in AD 1135. Scarcely thirty years later, in
1161, the Chin Tartars mounted another expedition against the Song, and
sought to make a crossing of the Yangtze. This led to the celebrated Battle
of Caishi, where, after many anxious moments, the Song forces gained the
day. Here the paddle-wheel warships repeated their achievements on the
river. Cruising rapidly around Jinshan Island, they constantly let off their
trebuchet artillery, and struck great fear into the hearts of the Jurchens, who
were not much accustomed to any kind of ships and found these almost
supernatural.

All through the Southern Song, there was great activity in building and
employing paddle-wheel ships. In AD 1134 warships with nine wheels and
thirteen wheels were built in the coastal provinces, and in 1183 a Nanjing
naval commander was specially rewarded for his work in building ninety
paddle-wheel and other ships. The imperial court itself took great interest
in its automotive vessels, unknown in any contemporary culture. Moreover,
in 1168, Admiral Shi Zhengzhi reported that he had constructed very eco-
nomically a 102 tonne warship propelled by a single twelve-bladed wheel.

This solves a problem. Evidently Shi Zhengzhi's paddle-boat must have
been a stern-wheeler, and the descriptions of boats with an odd number of
wheels must have had one stern wheel and an even number of pairs of side
wheels. We do not of course know details of the mechanisms, but if the
paddle-wheels of these Song designs were independently mounted, then
forward action on one side combined with reverse action on the other would
have rendered them eminently manoeuvrable – and some of the descriptions
emphasise this. Such an arrangement might have made a rudder unnecessary, and that would have been convenient if one of the large stern wheels
was fitted.

The power available was sometimes large. Thus, in AD 1203, two four-
wheeled 'sea-hawk' warships were constructed; they were covered over on
top, armoured with iron plates on their sides and fitted with spade-shaped
rams. The smaller, of some 100 tonnes burden, needed a propulsion crew
of twenty-eight men, the larger, of some 250 tonnes, required forty-two.
The largest numbers mentioned for a single ship is 200, but we do not know
whether this included relays of men. However, as is seen in Fig. 525, the
nineteenth century AD Chinese paddle-wheel passenger boats were worked by
three treadmills using coupling rods and eccentrics to drive the stern wheel.
Remembering the use of such devices in the hydraulic blowing-engines (see
p. 287) of the late thirteenth century AD, and allied machinery such as silk-

Fig. 525. Stern view of a Cantonese stern-wheel treadmill paddle-boat in the Museum f. Völkerkunde (Horwitz (1930)). The rear hull bulwarks have been removed so as to show the three eccentrics and coupling-rods like those of a steam locomotive; these were set 90° apart on both (port and starboard) sides to prevent dead-centre stoppages.

reeling and winding in the eleventh, it may well be that the Southern Song engineers made use of such methods.

So far, the accent has been on the naval use of paddle-wheel ships, but it seems probable also that smaller paddle-boats were used in the great Chinese harbours during the twelfth and thirteenth centuries AD, especially for towing. Indeed, we hear about them from the author of the *Meng Liang Lu* (The Past seems a Dream (description of the capital Hangzhou)) of 1275, who says

> There are also the wheel boats (such as those) belonging to the great house of Jia Qiuhe. On the deck above the cabin there are no men poling or rowing, for these craft move by means of wheels worked by a treadmill, and speed over the water like flying things.

Yet taking the evidence as a whole, one is struck by the fact that the development of paddle-boats in medieval China was primarily associated with their value in sea-fights, especially on lakes and rivers. The description in the *Imperial Encyclopaedia* (AD 1726) already noted, refers specifically to paddle-boats being used for naval purposes. The illustration (Fig. 523) is a century earlier than this, being derived from the *Treatise on Armament Technology* of 1628. The accompanying text is also assuredly much earlier than the eighteenth century AD, as in the case of some of the preceding warship illustrations, where the encyclopaedia reproduces almost verbatim an eighth century AD text. This one is indeed not so old as that, but it can be dated from intrinsic evidence. The passage describes the paddle-wheel barque, 12.8 metres long and 4 metres broad, having four treadmill-operated paddle-wheels dipping 30 centimetres into the water. It then considers the weapons used. There were *shen pao* [*shen phao*] (bombs or grenades), *shen jian* [*shen chien*] (incendiary arrows or rockets or incendiary rockets) and *shen huo* (spattering fire from fire-lances containing rocket composition, or perhaps burning petroleum like Greek Fire).

Such an account is certainly not from the Thang – the time of Li Gao – as are the other descriptions, though at first sight it might be Song, for toxic smoke bombs containing gunpowder are described in AD 1044 and are mentioned in many battle accounts of that period. But this term continued in use long afterwards, and the early Ming date of the text is more forcibly asserted by the recurring use of the word *shen*, 'magical', for three of the types of weapons. This was fashionable for all new inventions of this kind from about AD 1385 onwards. It seems therefore that we may reasonably date these paddle-boats as fifteenth century AD, for if the passage were of 1628 there would be more mention of barrel-guns and muskets and less emphasis on the earlier types of gunpowder weapons.

The persistence of the paddle-boat under the Ming is particularly interesting since it seems to have fallen so much out of the picture during the Yuan. Yet the Mongol dynasty did not neglect sea-power, but emphasised it so much that vessels for lake and river combats, like the paddle-boats, suffered a decline. Not until the coming of an adequate source of power were they suitable at sea.

Unlike some other medieval Chinese inventions, the treadmill paddle-boats survived in active use down to our own time, especially on the Pearl River at Canton (Fig. 526). Certainly, neither eccentrics nor connecting-rods can be seen, as in Fig. 525, but this merely suggests that a chain-drive replaced them. It is also known that the treadmill pedals were like those of the square-pallet chain-pump (Fig. 500). In the 1890s there were about fourteen of these stern-wheeler ships carrying some seventy passengers in large and roomy accommodation between Shanghai and Suzhou. Pedalled by a group of men six to twenty in number, they did the 160-kilometre journey in about a day and a night, thus averaging some 3.5 knots. We can thus understand the presence of these Mississippi-like boats in the Chinese rivers as late as the present century. Yet they had nothing to do either with the first European steamboats or with the Mississippi; they were in the direct line of descent from Admiral Shi Zhengzhi's stern-wheel battleship of AD 1168.

Even so, it is a remarkable fact that when modern Europeans first came to Chinese coastal waters they were quite unable to believe that such paddle-wheel boats could be anything more than an imitation of their own steam-boats. Yet Chinese scholars had been worrying about the history of the invention long before paddle-steamers ever appeared off the Chinese coasts. That remarkable man Fang Yizhi, mathematician, scientific encyclopaedist and finally Buddhist monk, has an entry on the subject in his *Wu Li Xiao Shi* (Small Encyclopaedia of the Principles of Things) of AD 1644, which reads:

> Foreign ships have a straight timber beneath (i.e. a keel) and are ballasted so as to be heavy below. They use wheels – but this was done of old.

He then goes on to describe the Han 'flying' wheel-boats with eight-bladed paddle-wheels and other evidence, which has already been mentioned. But the reference to foreign paddle-wheel boats is curious since his remarks were written nearly two centuries before any European boats of this kind could have been seen in Chinese waters. Possibly Fang confused propelling paddle-wheels with the wheels of chain-pumps used for pumping out the bilges. However, it seems more likely that he had news, either from Western sailors or Jesuit missionaries, of the experiments with treadmill-operated paddle-wheels which had been going on since the time of Blasco de Garay a century

Fig. 526. One of the treadmill paddle-boats of the Pearl River estuary near Canton. The large iron stern wheel, about 2.7 metres in diameter, can be seen fitted in the after- (rear) gallery under the steersman. Just in front of it, under the awning, there are three sets of handlebars for the pedallers, two rows of whom can be seen at work, resembling those who turn square-pallet chain-pumps (see Fig. 501). The funnel belongs to the white Japanese steam launch behind. (Photo. Paris, 1929.)

or so earlier. This in itself would be an interesting example of culture-contact.

The moment has now come to confront all these pieces of evidence. We have surely no option but to believe that the ship-mills of Belisarius (p. 303) were a direct consequence of the undershot water-wheels of Vitruvius. Once established in the middle of the sixth century AD, they continued their career in western Europe down to the present time. The origin of the paddle-wheel boat idea – 'ad-aqueous' water-wheels driven to cause motion – comes very much later in Europe (fifteenth century AD), in fact no sooner than the idea of making such boats wind themselves upstream. In China, on the other hand, the paddle-wheel boat is remarkably early. If our oldest references are faulty, and Li Gao is the true inventor, then it would be a plausible sugges-tion to propose that wandering Persian merchants in Tang China brought the simple message: 'In the West men have seen boats with wheels', upon which the Chinese assumed (wrongly) that ad-aqueous wheels were meant, and proceeded to construct them. On the other hand, if the real inventor was Zu Chongzhi or Wang Zhene in the fifth century AD, they cannot have been inspired by Belisarius, and their work may well have been a spontane-ous development of the (Indian?) noria, which would by that time have reached China. If substantiated, this would give us a possible date for the first introduction of vertical water-wheels of all kinds there. There is, how-ever, one further card in the pack which we have not yet played. It is the *gua che*, or scoop-wheel, already mentioned on p. 259, which is used for raising water where very small lifts were required, as from one field to another (15 centimetres or so). This wheel is ad-aqueous, and if indeed it was the first Chinese adaption of the ex-aqueous water-driven noria (for pushing rather than hoisting water), it might have put the idea into people's heads that something worthwhile might be performed by applying force *to* a paddle-wheel instead of using it to derive force *from* the water. Unfortu-nately, no literary references to the scoop-wheel have been found, apart from its mention in technical agricultural treatises. As for the origin of ship-mills in China, there is nothing definite to be said; they may have been an independent invention, or a quite separate introduction from Arab contacts, or even possibly a secondary adaption deriving from the paddle-wheel naval boats. We lack adequate information about them from pre-Song times.

Did the medieval paddle-boats have anything to do, one may wonder, with the literature on the 'Magic Boat', the self-moving ship on which voyagers saw neither sails nor machinery? Such descriptions are common in Arthurian legend, and in about AD 1100 in Irish tales, but they go back in stories of saints to around AD 690, at which time it was thought that auto-matic motion was caused by the presence of the saint or his relics. Three Coptic examples take this back even earlier, to the beginning of the seventh

Fig. 527. The *liburna*, or paddle-wheel ship, proposed in the manuscript *On Warlike Matters* written close to AD 370. (From Thompson and Flower (1952).) Six paddle-wheels are powered by three ox whims within the hull.

century AD. Could it be that the boat without crew is a far-away echo of the hidden pedallers of Zu Chongzhi and Wang Zhene, testifying to the superiority of Chinese technique in the fifth century AD, in contrast to its reversed position in the nineteenth.

But we still have not reached the dénoument. An anonymous manuscript *De Rebus Bellicis* (On Warlike Matters) had been thought by some scholars to be a forgery from the fourteenth century AD rather than being of the sixth century as others maintained. Nevertheless, by the 1960s, research had established that not only is it genuine, but that it was written close to the neighbourhood of AD 370. At some time – probably early seventh century AD – it was combined with several other Byzantine tracts. Then, in the ninth or tenth century, it was set down in a manuscript which became known as the Speyer Codex, only one leaf of which has survived to provide a dating. The significance for us is that *De Rebus Bellicis* illustrates and describes a ship of Roman type – a *liburna* – bearing three pairs of paddle-wheels turned by six oxen on the deck (Fig. 527), and there seems no likelihood that this *liburna* was an insertion from the fourteenth century, a time when the first copies of it were made.

For the origins of the anonymous author's idea, he was doubtless as much indebted to the vertical Vitruvian water-mill as to Belisarius later on. But as to its effects, scholars are all agreed that there is no contemporary mention of the idea, no evidence that it was anything more than a scheme on paper only. The whole text was, it is thought, probably intercepted by some civil servant and pigeon-holed without ever reaching the emperor to whom it was addressed; it even appears to have stayed in the files for half a millennium after it was written. In these peculiar circumstances it seems extremely unlikely that any word of the invention could have reached Zu Chongzhi or Wang Zhene a bare century later at the other end of the Old World. Li Gao's paddle-boats, in the latter half of the eighth century AD, were being built only fifty years or so after the manuscript's reappearance from obscurity, so that here again the possibilities of transmission seem small.

One cannot help feeling, then, that this constitutes the clearest evidence so far unravelled of a strong probability that essentially the same invention was made twice over in different places. Provisionally we can only say that the first specification was Byzantine and the first execution Chinese.

BIBLIOGRAPHY

Aldred, C. 'Furniture, to the end of the Roman Empire'. In *A History of Technology*, eds. C. Singer *et al.* Oxford, 1956, vol. 2, p. 220.

Alex, W. *Japanese Architecture*. Prentice-Hall, London; Braziller, New York, 1963.

Alley, R. 'Pagodas and towers in China'. *Eastern Horizon* (Hongkong), 1962, 2 (no. 5), 20.

Allom T. & Wright, G. N. *China, in a Series of Views, displaying the Scenery, Architecture and Social Habits of that Ancient Empire, drawn from original and authentic Sketches by T.A—Esq., with historical and descriptive Notices by Rev. G.N.W—.* 4 vols. Fisher, London & Paris, 1843.

Anon. [perhaps Capt. Kellett]. *Description of the Junk 'Keying', printed for the Author, and Sold on Board the Junk.* Such, London, 1848.

Anon. (tr.). 'The Chinaman abroad; or, a desultory account of the Malayan Archipelago, particularly of Java, by Ong-Tae-Hae' (Wang Dahai's *Hai Dao Yi Zhi Chai Lüe* of AD 1791).

Anon. *La Chine à Terre et on Ballon; Reproduction des Photographies des officiers du Génie du Corps Expéditionnaire, 1900–1901* (album). Berger-Levrault, Paris n.d. (1902?).

Anon. 'Miracle on the Yangtze' (an account of the Jing River retention-basin north of the Dongting Lake, with its three long regulator-sluice dams). *China Reconstructs*, 1952, 1 (no. 5), 6.

Anon. (ed.) *New China* (album of photographs). Foreign Languages Press, Peking, 1953.

Anon. 'The days of our years; has the time come to change our present calendar?' *Unesco Courier*, 1954, 7 (no. 1), 28

Anon. 'A canal through the mountains' (the utilisation of the water of the Thao River in Southern Gansu). *Peking Review*, 1958 (no. 24), 17.

Bannister, T. C. 'The first iron-framed buildings'. *Architectural Review*, 1950, **107**, 231.

Barrows, H. K. *Floods; their Hydrology and Control.* McGraw-Hill, New York, 1948.

Beaton, C. *Chinese Album* (photographs). Batsford, London, 1945.

de Bélidor, B. F. *Architecture Hydraulique; ou l'Art de Conduire, d'Elever et de Menager les Eaux, pour les différens Besoins de la Vie.* 4 vols. Jombert, Paris, 1737–53.

Bishop, C. W. 'An ancient Chinese capital; earthworks at Old Chang'an.' *Antiquity.* 1938, **12**, 68.

Boerschmann, E. *Baukunst und Religiöse Kultur der Chinesen.* 2 vols.; vol. 1, Pu Tuo Shan (the famous island with its many Buddhist temples off the coast of Jiangsu); vol. 2, Gedächtnistempel (memorial temples both Daoist and Confucian, esp. those of Zhang Liang at Miaotai in Shaanxi of Li Bing and Li Erlang at Guanxian in Sichuan, of Confucius in Shandong, etc. etc.). Reinier, Berlin, 1911.

Boerschmann, E. *China; Architecture and Landscape – a Journey through Twelve Provinces.* Studio, London, n.d. (1928–9).

Booker, P. J. *A History of Engineering Drawing.* Chatto & Windus, London, 1963. Rev. D. Chilton, *Technology and Culture,* 1965, **6**, 128.

Boyd, Andrew. *Chinese Architecture* (Introduction to the Catalogue of the Exhibition prepared by the Architectural Society of China and shown at the Royal Institute of British Architects, 1959). RIBA, London, 1959.

van Braam Houckgeest, A. E. *An Authentic Account of the Embassy of the Dutch East-India Company to the Court of the Emperor of China in the years 1794 and 1795 (subsequent to that of the Earl of Macartney), containing a Description of Several Parts of the Chinese Empire unknown to Europeans; taken from the Journal of André Everard van Braam, Chief of the Direction of that Company, and Second in the Embassy.* Tr. L. E. Moreau de St Méry. 2 vols., map, but no index and no plates; Phillips, London, 1798. French edn 2 vols., with map, index and several plates; Philadelphia, 1797. The two volumes of the English edition correspond to vol. 1 of the French edition only.

Briggs, M. S. 'Building construction [in the Mediterranean civilisations and the Middle Ages]'. In *A History of Technology,* eds. C. Singer *et al.* Oxford, 1956, vol. 2, p. 397.

Brohier, R. L. *Ancient Irrigation Works in Ceylon.* 3 vols. Ceylon Government Press, Colombo, 1934–5. Vol. 1 reprinted 1939; vol. 2 reprinted 1940.

Carles, W. R. 'The Grand Canal of China.' *Journal (or Transactions) of the North China Branch of the Royal Asiatic Society,* 1896, **31**, 102.

Cescinsky, H. *Chinese Furniture; a series of Examples from the Collections in France.* London, 1922.

Chambers, Sir W. *Designs of Chinese Buildings, Furniture, Dresses, Machines and Utensils; to which is annexed, A Description of their Temples, Houses, Gardens, etc.* London, 1757.

Chatley, H. 'The hydrology of the Yangtze River.' *Journal of the Institute of Civil Engineers (UK),* 1939, pp. 227 and 565 (Paper no. 5223).

Chatley, H. 'The Yellow River as a factor in the development of China.' *Asiatic Review,* 1939, 1.

Chatley, H. 'Far Eastern engineering.' *Transactions of the Newcomen Society,* 1954, **29**, 151. With discussion by J. Needham, A. Stowers, A. W. Skempton, S. B. Hamilton *et al.*

Chavannes, E. *La Sculpture sur Pierre en Chine aux Temps des deux dynasties Han.* Leroux, Paris, 1893.

Chavannes, E. 'L'instruction d'un futur Empereur de Chine en l'an 1193' [on the astronomical, geographical, and historical charts inscribed on stone steles in the Confucian temple at Suzhou, Jiangsu]. In *Mémoires concernant l'Asie Orientale* (publ. Acad. des Inscriptions et Belles Lettres), Leroux, Paris, 1913, vol. 1, p. 19.

Chin Shou-Shen. *See* Jin Shousen.

Chu Chhi-Chhien & Yeh Kung-Chao. *See* Zhu Qiqian & Ye Gongzhao.

Cross, H. & Freeman, J. R. (eds.). *River Control and the Yellow River of China; a Collection of the Opinions of China.* 2 vols. Brown Univ., Providence, RI, 1918.

Davey, N. *A History of Building Materials.* Phoenix, London, 1961.

Davies, R. M. *Yunnan, the Link between India and the Yangtze.* Cambridge, 1909.

Dawson, R. (ed.). *The Legacy of China.* Oxford, 1964.

Drower, M. S. 'Water-supply, irrigation and agriculture [from early times to the end of the ancient empires]'. In *A History of Technology,* eds. C. Singer *et al.* Oxford, 1954, vol. 1, p. 520.

Dukes, E. J. *Everyday Life in China; or Scenes along River and Road in Fukien.* London, 1885, pp. 144–50. (Gives an account of Cai Xiang and the building of the Luoyang or Wan-an megalithic beam bridge at Quanzhou.)

Ecke, G. V. 'Chiang Tung Chhiao; eine Brücke in Sud-Fukien aus der Zeit d. Nan Sung.' *Ostasiatische Zeitschrift,* 1929, **15,** 110.

Edgar, J. H. 'From Ta-tsien-lu to Mu-phing via Yü-thung.' *China Journal of Science and Arts,* 1932, **17,** 282.

Ewbank, T. *A Descriptive and Historical Account of Hydraulic and other Machines for Raising Water, Ancient and Modern. . . .* Scribner, New York, 1842. (Best edition is the 16th, 1870.)

Farrer, R. (2). *The Rainbow Bridge.* Arnold, London, 1926.

Fisher, B. 'The *qanāts* of Persia.' *Geographical Review,* 1928, **18,** 302.

Fitzgerald, C. P. *Son of Heaven; a Biography of Li Shih-Min, Founder of the Thang Dynasty.* Cambridge, 1933.

Forbes, R. J. *Notes on the History of Ancient Roads and their Construction.* A. P. Stichting. Amsterdam, 1934; Brill, Leiden, 1934. [Archaeologische-Historische Bijdragen d. Allard Pierson Stichting, Amsterdam, no. 3.]

Forbes, R. J. *Studies in Ancient Technology. Vol. 2, Irrigation and Drainage; Power; Land Transport and Road-Building; The Coming of the Camel.* Brill, Leiden, 1955.

Forman, W. & Forman, B. *Das Drachenboot* (album of photographs of Chinese places, buildings, vessels, etc.). Artia, Prague, 1960.

Fugl-Meyer, H. *Chinese Bridges.* Kelly & Walsh, Shanghai, 1937.

Gallagher, L. J. (tr.) *China in the 16th Century; the Journals of Matthew Ricci, 1583–1610.* Random House, New York, 1953. (A complete translation, preceded by inadequate bibliographical details, of Nicholas Trigault's *De Christiana Expeditione apud Sinas* (1615). Based on an earlier publication: *The China that Was; China as discovered by the Jesuits at the close of the 16th Century: from the Latin of Nicholas Trigault.* Milwaukee, 1942.)

Geil, W. E. *The Great Wall of China.* Murray, London, 1909.

Geil, W. E. *The Eighteen Capitals of China.* Constable, London, 1911.

Gill, W. *The River of Golden Sand, being the narrative of a Journey through China and Eastern Tibet to Burmah,* eds. E. C. Baber & H. Yule. Murray, London, 1883.

Goodchild, R. G. & Forbes, R. J. 'Roads and land travel [including bridges, cuttings, tunnels, harbours, docks and lighthouses] (in the Mediterranean civilisations and the Middle Ages)'. In *A History of Technology,* eds. C. Singer *et al.* Oxford, 1956, vol. 2, p. 493.

Goodrich, L. Carrington. 'Suspension-bridges in China.' *Sino-Indian Studies* (Santiniketan), 1957, **5**, (nos. 3–4), 1.

Gregory, R. 'How the eyes deceive'. *Listener* (BBC), 1962, **68** (no. 1736), 15.

Hahnloser, H. R. (ed.). *The Album of Villard de Honnecourt.* Schroll, Vienna, 1935.

Hart, I. B. *The World of Leonardo da Vinci, Man of Science, and Dreamer of Flight.* McDonald, London, 1961.

Herrmann, A. *Historical and Commerical Atlas of China.* Harvard-Yenching Institute, Cambridge, MA, 1935.

Horwitz, H. T. 'Zur Geschichte des Schaufelradtriebes.' *Zeitschrift d. österr. Ingenieur u. Architekten Vereines*, 1930, **82**, 309 & 356.

Hsüeh Pei-Yuan. *See* Xue Beiyuan.

Hummel, A. W. 'History of the Kuangling iron suspension-bridge [over the northern Phan Chiang in Kweichow].' *Quarterly Journal of Current Acquisitions* (Library of Congress, Washington), 1948, **5**, 23.

Jin Shousen. 'The Great Wall of China.' *China Reconstructs*, 1962, **11** (no. 1), 20.

Karlgren, B. (tr.) *The Book of Odes; Chinese Text, Transcription and Translation.* Museum of Far Eastern Antiquities, Stockholm, 1950.

King, F. H. *Irrigation in Humid Climates.* US Dept. of Agriculture Farmers' Bull, no. 46, Government Printing Office Washington, D.C., 1896.

Krenkow, F. 'The construction of subterranean water supplies during the Abbasid Caliphate'. *Transactions of the Glasgow University Oriental Society*, 1951, **13**, 23.

Latham, R. E. (ed.). *The Travels of Marco Polo.* Penguin, London, 1958.

Lattimore, O., Royd O., Lord, Robinson, J., Needham, J. & Keswick, J. 'Chi Chhao-Ting – scholar revolutionary.' *Arts and Sciences in China* (London), 1964, **2** (no. 1), 9.

Laufer, B. *Chinese Pottery of the Han Dynasty.* (Publication of the East Asiatic Committee of the American Museum of Natural History), Brill, Leiden, 1909. (Photolitho re-issue, Tientsin, 1940.)

Laufer, B. 'Christian art in China.' *Mitteilungen d. Seminar f. Orientalischen Sprachen* (Berlin), 1910, **13**, 100.

Laufer, B. 'The noria or Persian wheel'. *Art in Oriental Studies in honour of Cursetji Erachji Pavry*, ed. A. V. W. Jackson, Oxford, 1933, p. 238.

Lecomte, Louis. *Nouveaux Mémoires sur l'État présent de la Chine.* Anisson, Paris, 1696. (Eng. tr. *Memoirs and Observations Topographical, Physical, Mathematical, Mechanical, Natural, Civil and Ecclesiastical, made in a late journey through the Empire of China, and published in several letters, particularly upon the Chinese Pottery and Varnishing, the Silk and other Manufactures, the Pearl Fishing, the History of Plants and Animals, etc. translated from the Paris edition, etc.* 2nd edn, London, 1698.)

Lee, S. (tr.) *The Travels of Ibn Baṭṭuṭah.* Oriental Translation Committee, Royal Asiatic Society, London, 1829.

Liang Sicheng. 'China's architectural heritage and the tasks of today'. *People's China*, 1952 (Nov.), 30.

Liang Ssu-Chheng. *See* Liang Sicheng.

Lo Jung Pang. *See* Lo Rongbang.

Lo Rongbang. 'China's paddle-wheel boats; the mechanised craft used in the opium war and their historical background'. *Ch'ing-Hua (T'sing-Hua) Journal of Chinese Studies* (new series, publ. Taiwan), 1960 (n.s.), **2** (no. 1), p. 189.

Lu Gwei-Djen. 'China's greatest naturalist; a brief biography of Li Shih-Chen'. *Physis* (Florence), 1966, **8**, 383. Abridgement in *Proc. XIth Internat. Congress of the History of Science*, Warsaw, 1965, vol. 5, p. 50.

McGregor, J. 'On the paddle-wheel and screw propeller, from the earliest times'. *Journal of the Royal Society of Arts*, 1858, **6**, 335.

Mao I-Shêng. *See* Mao Yisheng.

Mao Yisheng. 'The stone arch – symbol of Chinese bridges'. *China Reconstructs*, 1961, **10** (no. 11), 18.

March, B. *Some Technical Terms of Chinese Painting.* American Council of Learned Societies, Waverly, Baltimore, 1935. (ACLS Studies in Chinese and Related Civilisations, no. 2.)

de Mendoza, Juan Gonzales. *Historia de las Cosas mas notables, Ritos y Costumbres del Gran Reyno de la China, sabidas assi por los libros de los mesmos Chinas, como por relacion de religiosos y oltras personas que an estado en el dicho Reyno.* Rome, 1585 (in Spanish). Eng. tr. Robert Parke, *The Historie of the Great & Mightie Kingdome of China and the Situation thereof; Togither with the Great Riches, Huge Citties, Politike Gouvernement and Rare Inventions in the same* [undertaken 'at the earnest request and encouragement of my worshipfull friend Master Richard Hakluyt, late of Oxforde']. London, 1588 (1589). New edn G. T. Staunton, London, 1853 (Hakluyt Society Pubs., 1st series, nos. 14, 15).

Mirams, D. G. *A Brief History of Chinese Architecture.* Kelly & Walsh, Shanghai, 1940.

Molenaar, A. *Water Lifting Devices for Irrigation.* FAO, Rome, 1956. (Agricultural Development Paper, no. 60.)

Moule, A. C. 'The bore on the Ch'ien-T'ang River in China'. *T'oung Pau (Archives concernant l'Histoire, les Langues, la Géographie, l'Ethnographie et les Arts de l'Asie Orientale*, Leiden), 1923, **22**, 135 (includes much material on tides and tidal theory).

Moule, A. C. & Pelliot, P. (tr. and annot.). *Marco Polo (AD 1254 to 1325); The Description of the World.* 2 vols. Routledge, London, 1938. Further notes by P. Pelliot (posthumously pub.), 2 vols. Impr. Nat., Paris, 1960.

Needham, J. *The Development of Iron and Steel Technology in China.* Newcomen Society, London, 1958. (Second Biennial Dickinson Memorial Lecture, Newcomen Society.) Reprinted by Heffer, Cambridge, 1964.

Needham, J. 'China and the invention of the pound-lock'. *Transactions of the Newcomen Society*, 1964, **36**, 85.

Nieuhoff, J. *L'Ambassade [1655–7] de la Compagnie Orientale des Provinces Unies vers l'Empereur de la Chine, ou Grand Cam de Tartarie, faite par les Sieurs Pierre de Goyer & Jacob de Keyser; Illustrée d'une tres-exacte Description des Villes, Bourgs, Villages, Ports de Mers, et autres Lieux plus considerables de la Chine; Enrichie d'un grand nombre de Tailles douces, le tout receuilli par Mr Jean Nieuhoff. . .* (title of Pt. II: *Description Generale de l'Empire de la Chine, ou il est traité succinctement du Gouvernement, de la Religion, des Mœurs, des Sciences et Arts des Chinois, comme aussi des Animaux, des Poissons, des Arbres et Plantes, qui ornent leurs Campagnes et leurs Rivieres; y joint un court Recit des dernieres Guerres qu'ils ont eu contre les Tartares).* de Meurs, Leiden, 1665.

Parson, A. W. 'A Roman water-mill in the Athenian Agora'. *Hesperia (Journal of the American School of Classical Studies Athens)*, 1936, **5**, 10.

Playfair, G. M. H. 'The grain transport system of China; notes and statistics taken from the *Ta Chhing Hui Tien*'. *China Review* (Hongkong and Shanghai), 1875, **3**, 354. [(1) The personnel of the transport service; (2) The itinerary of the Grand Canal; (3) Tribute; (4) White rice tribute; (5) The building and repairing of junks; (6) Grain fleets.]

Prip-Moller, J. *Chinese Buddhist Monasteries; their Plan and Function as a Setting for Buddhist Monastic Life.* Oxford, 1937. Reprinted with biographical and bibliographical notes, Vetch, Hongkong, 1968.

Pugsley, Sir Alfred. *The Theory of Suspension Bridges.* Arnold, London, 1957.

Ramsay, A. M. 'The speed of the Roman Imperial Post'. *Journal of Roman Studies*, 1925, **15**, 73.

Reischauer, E. O. (tr.). *Ennin's Diary; the Record of a Pilgrimage to China in Search of the Law* (the *Nitto Guho Junrei Gyōki*). Ronald Press, New York, 1955.

Reischauer, E. O. *Ennin's Travels in Thang China.* Ronald Press, New York, 1955.

Richter, G. M. A. *The Furniture of the Greeks, Etruscans and Romans.* Phaidon, London, 1966.

Robins, F. W. *The Story of Water Supply.* Oxford, 1946.

Robins, F. W. *The Story of the Bridge.* Cornish, Birmingham, n.d. (1948).

Rock, J. F. *The Ancient Na-Khi Kingdom of Southwest China.* 2 vols. (with magnificent collotype illustrations). Harvard University Press, Cambridge, MA, 1947. (Harvard-Yenching Monograph Series, no. 9.)

Rose, A. *Public Roads of the Past.* Washington, D.C., 1952.

Schnitter, N. J. 'A short history of dam engineering'. *Water Power*, 1967, **19** (no. 4), 142.

Schreiber, H. *The History of Roads.* Barrie & Rockliff, London, 1962.

Scott, John. *The Complete Text-book of Farm Engineering; comprising Practical Treatises on Draining and Embanking, Irrigation and Water-Supply, Farm Roads, Fences and Gates, Farm Buildings, Barn Implements and Machines, and Agricultural Surveying.* Crosby Lockwood, London, 1885.

Seaton, A. E. *The Screw Propeller, and other Competing Instruments for Marine Propulsion.* Griffin, London, 1909.

Shadick, H. (tr. & ed.). '*The Travels of Lao Tshan*', by Liu Thieh-Yün (Liu É). Cornell University Press, Ithaca, NY, 1952.

Sickman, L., Loehr, M., Yang Lien-Shêng & Sullivan, M. *Chinese Painting and Calligraphy from the Collection of John M. Crawford Jr. [Catalogue of an Exhibition with Introductions].* Arts Council of Gt. Britain, Victoria and Albert Museum, London, 1965.

Sickman, L. & Soper, A. *The Art and Architecture of China.* Penguin (Pelican), London, 1956. (Rev. A. Lippe, *Journal of the Asiatic Society*, 1956, **11**, 137.) New edn 1968.

Sirén, O. *The Walls and Gates of Peking.* London, 1924.

Sirén, O. (10). *Chinese Painting; Leading Masters and Principles.* Lund Humphries,

London, 1956; Ronald, New York, 1956. 7 vols. Pt. 1, The First Millennium, 3 vols. incl. one of plates; pt. 11, The Later Centuries, 4 vols., incl. one of plates.

Smythe, F. S. 'Suspension bridges on the Nepal-Tibet border'. *Geographical Magazine*, 1938, 7, 189.

Staunton, Sir George Leonard. *An Authentic Account of an Embassy from the King of Great Britain to the Emperor of China . . . taken chiefly from the Papers of H.E. the Earl of Macartney, K.B. etc. . . .* 2 vols. Bulmer & Nicol, London, 1797; reprinted 1798. Abridged Eng. edn, 1 vol. Stockdale, London, 1797.

Steward, Julian (ed.). *Irrigation Systems*. Washington, D.C., 1956. (Pan-American Union, Social Science Monographs, no. 1.)

Stone, L. H. *The Chair in China*. Royal Ontario Museum of Archaeology, Toronto, 1952.

Stowers, A. 'Observations on the history of water-power'. *Transactions of the Newcomen Society*, 1960, 30, 239.

Sullivan, M. *An Introduction to Chinese Art*. Faber & Faber, London, 1961.

Sullivan, M. 'The heritage of Chinese art'. In *The Legacy of China*, ed. R. Dawson, Oxford, 1964, p. 165.

Thompson, E. A. & Flower, B. *A Roman Reformer and Inventor; being a New Text of the Treatise 'De Rebus Bellicis', with a translation . . . introduction . . . and Latin index. . . .* Oxford, 1952. This text is now generally conceded to have been written by a Latin of Illyria in the close neighbourhood of AD 370.

Waley, A. *An Introduction to the Study of Chinese Painting*. Benn, London, 1923; reprinted 1958.

Waley, A. 'A Chinese picture' (Chang Tsê-Tuan's *Going up the River to Kaifêng at the Spring Festival, c.* AD 1126). *Burlington Magazine*, 1917, 30, 3.

Wang Pi-Wên. 'Official regulations of the Ch'ing Dynasty for the designing of locks and culverts'. *Bulletin of the Society for Research in [the History of] Chinese Architecture*, 1935, 6 (no. 2), 49.

Watson, Burton (tr.). '*Records of the Grand Historian of China*', translated from the '*Shih Chi*' of Ssuma Chhien. 2 vols. Columbia University Press, New York, 1961.

Watson, Burton. *Ssuma Chhien, Grand Historian of China*. New York, 1958.

Wells, W. H. *Perspective in Early Chinese Painting*. Goldston, London, 1935; reprinted 1945.

Westcott, G. F. *Pumping Machinery. Pt. 1. Historical Notes*. (Handbook of the Collections, Science Museum, South Kensington.) HMSO, London, 1932.

Wilson, P. N. 'The origins of water-power, with special reference to its use and economic importance in England from Saxon times to AD 1750.' *Water Power*, 1952, p. 308.

Wilson, P. N. *Watermills with Horizontal Wheels*. Wilson, Kendal, 1960. (Society for the Protection of Ancient Buildings, Wind and Watermill Section, Booklet series, no. 7.)

Wo Noson. *Chinese and Indian architecture*. Prentice-Hall, London, 1963; Braziller, New York, 1963.

Worcester, G. R. G. *Junks and Sampans of the Upper Yangtze*. Inspectorate-General of Customs, Shanghai, 1940. (China Maritime Customs Pub., ser. III, miscellaneous, no. 51.)

Wu No-Sun. *See* Wo Noson.

Wulff, H. E. 'The *qanāts* of Iran.' *Scientific American*, 1968, **218** (no. 4), 94.

Xue Beiyuan. 'Water conservancy two thousand years ago' (the Guanxian works, the Zhengguo Canal and the Qin irrigation canal along the Yellow River in Ningxia). *China Reconstructs*, 1957, **6** (no. 10), 9.

Zhu Qiqian & Ye Gongzhao. '[Chinese] Architecture; a brief Historical Account based on the Evolution of the City of Peiping.' In *Symposium on Chinese Culture*, ed. Sophia H. Chen, Zen. Inst. Pacific Relations, Shanghai, 1931, p. 97.

Zonca, Vittorio. *Novo Teatro di Machini e Edificii.* Bertelli, Padua, 1607 and 1621.

TABLE OF CHINESE DYNASTIES

夏 Xɪᴀ [Hꜱɪᴀ] kingdom (legendary?)		c. −2000 to c. −1520
高 Sʜᴀɴɢ (Yɪɴ) kingdom		c. −1520 to c. −1030
周 Zʜᴏᴜ [Cʜᴏᴜ] dynasty (Feudal Age)	Early Zhou [Chou] period	c. −1030 to −722
	Chun Qiu [Chhun Chhiu] period	−722 to −480
	Warring States (Zhan Guo [Chan Kuo]) period 戰國	−480 to −221
First Unification 秦 Qɪɴ [Cʜʜɪɴ] dynasty		−221 to −207
漢 Hᴀɴ dynasty	Qian Han [Chhien Han] (Earlier or Western)	−202 to +9
	Xin [Hsin] interregnum	+9 to +23
	Hou Han (Later or Eastern)	+25 to +220
三國 Sᴀɴ Gᴜᴏ [Sᴀɴ Kᴜᴏ] (Three Kingdoms period)		+221 to +265
First	蜀 Sʜᴜ (Hᴀɴ) +221 to +264	
Partition	魏 Wᴇɪ +220 to +265	
	吳 Wᴜ +222 to +280	
Second Unification	晉 Jɪɴ [Cʜɪɴ] dynasty: Western	+265 to +317
	Eastern	+317 to +420
	劉宋 (Liu) Sᴏɴɢ [Sᴜɴɢ] dynasty	+420 to +479
Second Partition	Northern and Southern Dynasties (Nan Bei chao [Nan Pei chhao])	
	齊 Qɪ [Cʜʜɪ] dynasty	+479 to +502
	梁 Lɪᴀɴɢ dynasty	+502 to +557
	陳 Cʜᴇɴ [Cʜʜᴇɴ] dynasty	+557 to +589
	魏 { Northern (Touba [Thopa]) Wᴇɪ dynasty	+386 to +535
	Western (Touba) Wᴇɪ dynasty	+535 to +556
	Eastern (Touba) Wᴇɪ dynasty	+534 to +550
	北齊 Northern Qɪ [Cʜʜɪ] dynasty	+550 to +577
	北周 Northern Zʜᴏᴜ [Cʜᴏᴜ] (Xienbi [Hsienpi]) dynasty	+557 to +581
Third Unification	隋 Sᴜɪ dynasty	+581 to +618
	唐 Tᴀɴɢ [Tʜᴀɴɢ] dynasty	+618 to +906
Third Partition	五代 Wᴜ Dᴀɪ [Wᴜ Tᴀɪ] (Five Dynasty period) (Later Liang, Later Tang [Thang] (Turkic), Later Jin [Chin] (Turkic), Later Han (Turkic), and Later Zhou [Chou])	+907 to +960
	遼 Lɪᴀᴏ (Qidan [Chhitan] Tartar) dynasty	+907 to +1124
	West Lɪᴀᴏ dynasty (Qarā-Khitāi)	+1124 to +1211
	西夏 Xi Xia [Hsi Hsia] (Tangut Tibetan) state	+986 to +1227
Fourth Unification	宋 Northern Sᴏɴɢ [Sᴜɴɢ] dynasty	+960 to +1126
	宋 Southern Sᴏɴɢ [Sᴜɴɢ] dynasty	+1127 to +1279
	Jɪɴ [Cʜɪɴ] (Jurchen Tartar) dynasty	+1115 to +1234
	元 Yᴜᴀɴ (Mongol) dynasty	+1260 to +1368
	明 Mɪɴɢ dynasty	+1368 to +1644
	清 Qɪɴɢ [Cʜʜɪɴɢ] (Manchu) dynasty	+1644 to +1911
	民國 Republic	+1912

N.B. When no modifying term in brackets is given, the dynasty was purely Chinese. During the Eastern Jin period there were no less than eighteen independent States (Hunnish, Tibetan, Xienbi, Turkic, etc.) in the north. The term 'Liu chao' [Liu chhao] (Six Dynasties) is often used by historians of literature. It refers to the south, and covers the period from the beginning of the third to the end of the sixth centuries AD, including (San Guo) Wu, Jin, (Liu) Song, Qi, Liang and Chen. The minus sign (−) indicates BC and the plus sign (+) is used for AD.

INDEX

Wang Shizhen [Wang Shih-chen]
(official), ship-mills, 304
Wang Wu, Bian Canal, 197
Wang Yanhui, paddle-wheel boats,
309
Wang Yinglin [Wang Ying-Lin]
(encyclopaedist), 206–8
Wang Zhen [Wang Chên]
(agriculturalist), hydraulic
engineering, 277, 289–90, 291,
292, 293, 297, 302
Wang Zhene [Wang Chen-O]
(admiral), paddle-wheel boats, 308,
318, 320
Wang Zhenpeng [Wang Chen-Phêng]
(painter), 166, 170
Wang Zhijian [Wang Chih-Chien]
(writer), 147
Wangdu, Han tower model, 95
*Wannian Qiao Zhi [Wan-Nien Chhiao
Chih]* (Record of the Bridge of
Ten Thousand Years), 141, 142–5
Wannian [Wan-Nien] Bridge, 141
Wanyan Liang [Wanyen Liang]
(emperor), paddle-wheel boats,
312
Warring States
architectural records, 62
building traditions, 89
Great Wall, 42, 43
Magic Transport Canal, 216
water as a weapon, 193
warships, paddle-wheel boats, 307,
308–13, 315–16, 319
Water Conservancy of the Wu District,
see *Wu Zhong Shui Li Shu*
water-mills, 296–302
sluices and gates, 241
water-wheels in the West, 285–7
weirs, 241
water-power, 282–303
metallurgical blowing-engines,
287–91, 302
spoon tilt-hammers, 282–4
spread of inventions, 302–3
steam engine ancestry, 291–3
trip-hammers, 291, 293–6, 297, 302,
303–4
water-mills, 296–302
water-wheels, 285–7

water-raising machinery, 255–82
counterbalanced bailing buckets,
255–9
norias (peripheral pot wheels), 270,
277–82
rag-and-chain pumps, 274–5
sāqīya (pot chain-pumps), 274, 275–7
scoop-wheels, 259, 262, 263, 318
square-pallet chain-pumps, 263–74
well-windlasses, 259, 260, 261
see also water-power
waterway transport, 173, 174, 297
Baoye Road and, 16
double-slipways, 250–1
Grand Canal, 183, 219, 221, 223, 227
locks, 176
Magic Transport Canal (Ling Qu),
212–13, 216–18
paddle-wheel boats, 306–20
pound-locks and, 246–7, 249–50
tax-grain system, 182–3
waterways, 172–254
Bian Canal reconstruction, 197
bridges, 128–9, 130, 132, 134, 140
centralisation of government and,
192–3
Chang'an–Yellow River Canal, 196–7
climatic factors, 177–82, 189, 253–4
drainage, 231–2
dredging, 232–3, 235, 274
dual function, 173–4
forests and, 187–9, 191
Grand Canal, see Grand Canal
Guanxian irrigation system, 190–1,
198, 202–10
Han Kou, 196
Hong Gou (Canal of the Wild
Geese), 193, 195–6, 216, 218
international achievements compared,
235, 241, 247, 248–9, 250, 251–4
for irrigation, see irrigation systems
Kunming reservoirs, 210–11
legends, 189–91
Ling Qu (Magic Transport Canal),
198, 212–18, 223
literature on, 229–30, 231
locks, see locks (waterways)
manpower, 192, 194, 195
methods for harnessing river waters,
176–7

Printed in the United States
by Bookmasters

Printed in the United States
By Bookmasters